CONSTRUCTION SCIENCE
PART THREE

CONSTRUCTION SCIENCE

PART THREE

Edwin Walker BSc and Selwyn Morgan BSc

HUTCHINSON EDUCATIONAL

HUTCHINSON EDUCATIONAL LTD
3 Fitzroy Square, London W1

London Melbourne Sydney Auckland
Wellington Johannesburg Cape Town
and agencies throughout the world

First published 1971

D
691
WAL

*This book has been set in Imprint type, printed in Great Britain
on smooth wove paper by Anchor Press, and
bound by Wm. Brendon, both of Tiptree, Essex*

ISBN 0 09 109780 0 (cased)
    0 09 109781 9 (paper)

Contents

Preface and Acknowledgements

Construction Science meets the science requirements of students taking an Ordinary National Certificate in Construction and is written in three volumes. The first part will deal with the first-year syllabus (01) while the second year (02) is covered by parts 2 (already published) and 3. S.I. units are used throughout.

A large number of practical exercises have been included in order to allow an adequate selection appropriate to the needs of particular groups of students. Worked examples are included in the text and graded questions and problems are to be found at the end of each chapter.

We also hope that the book will contain much information of use to craft students and technicians, and prove helpful to architects and engineers.

Data from, or based on, the British Standards listed below is reproduced by permission of the British Standards Institution, 2 Park Street, London W1Y 4AA, from whom copies of the complete publications may be obtained.

12:1958	Portland cement (ordinary and rapid hardening).
812:1967	Methods for sampling and testing of mineral aggregates, sands and fillers.
890:1966	Building limes.
1191:1967	Gypsum building plasters—excluding pre-mixed (part one) lightweight plasters.
1881:1970	Methods of testing concrete (metric edition).
C.P. 114	Structural use of reinforced concrete in buildings (metric edition).

Many of the practical exercises in this book are in close accord with the British Standard test methods since examination syllabuses frequently demand a general familiarity with them. They have, however, been slightly modified to meet the teaching laboratory situation and it must be stressed that the original publications should be consulted when carrying out tests to determine the compliance, or otherwise, of any of the materials. Where metric equivalents have been given in a British Standard using Imperial units these have been used, otherwise conversions have been made. It is desirable to consult the up-to-date edition of a standard since, even where the British Standards have been revised in terms of metric equivalents, this may only be a preliminary to the preparation of a proposed 'metricated' edition.

We are also grateful to the Road Research Laboratory for permission to use tables and graphs from Road Note 4, *Design of Concrete Mixes*, to Imperial Chemical Industries Ltd. for information included in the chapters on paints and plastics, and to International Nickel Ltd., whose film *Corrosion in Action* we find so valuable.

One Lime

Lime is manufactured from chalk or limestone which consist largely of calcium carbonate. Limestone occurs in the Midlands, in the North and West of England, and in Wales and Ireland, while chalk covers large areas of South-Eastern England. White chalk and mountain limestone are the purest material, containing over 95% calcium carbonate. Grey chalk, or greystone, contains silica and alumina as impurities which confer valuable properties to the lime.

The raw material is quarried, crushed, and heated, or *calcined* in specially designed kilns. Here the calcium carbonate decomposes with the evolution of carbon dioxide gas to yield calcium oxide, or quicklime, according to the following reaction:

$$CaCO_3 \xrightarrow[1400°C]{Heat,} CaO + CO_2$$

Commercially lime is available in several forms:

(a) *Quicklime*
Quicklime is marketed in a range of sizes graded to suit individual requirements. The *lump* quicklime may be *run-of-the-kiln*, indicating that it is as drawn from the kiln, or *hand-picked*, indicating that any overburnt or underburnt lumps, flints, and other unwanted material have been removed. *Ground quicklime* has been reduced to a fine powder.

Quicklime is used for the preparation of ready-mixed mortars.

(b) *Hydrated lime*
When water is added to quicklime, a process known as 'slaking', a considerable amount of heat is generated in a chemical reaction which produces calcium hydroxide, or hydrated lime—

$$CaO + H_2O \longrightarrow Ca(OH)_2 + Heat$$

The water addition is carefully controlled so that excess is driven off as steam by the heat of reaction to produce hydrated lime as a fine dry powder.

Hydrated lime is used for ready mixed mortars and for mixing on the site.

(c) *Lime putty*

When quicklime is slaked with an excess of water and allowed to mature lime putty is produced.

Lime putty is used where mixing is done on the site.

Setting of lime

When lime is used it is first converted into putty lime if it is not obtained in this form. After application putty lime sets by the evaporation of water; no chemical reaction occurs during this stage—it is solely a drying-out process.

Hardening of lime

After the lime has set it hardens by the absorption of carbon dioxide from the atmosphere to re-form calcium carbonate—

$$Ca(OH)_2 + CO_2 \longrightarrow CaCO_3 + H_2O$$

The water produced evaporates as the reaction proceeds.

The hardening is a gradual process requiring the presence of moisture; consequently if the lime is dried out too quickly the carbonation will be incomplete and a low strength will result. Since the surface of the lime absorbs the carbon dioxide first a hard skin is formed which will retard the carbonation of the interior. This means that such a plaster must be applied in several coats, building up to the required thickness. A single thick coat would result in a soft weak interior covered by a hard skin.

As the lime dries out it consolidates to make good the voids left by the evaporating water. This results in shrinkage and cracking. In order to reduce the shrinkage and to assist the penetration of carbon dioxide into the centre of the plaster the lime is mixed with sand in a traditional 3-coat process:

(a) *Backing coat*
3 parts sand
1 part putty lime
Hair

This coat was applied to a thickness of 6 to 9 mm and allowed to carbonate for several weeks.

(b) *Floating coat*
The composition of this coat was generally similar to the backing coat. The suction of the latter causes the floating coat to stiffen and allows it to be consolidated with the hand float to reduce cracking.

(c) *Finishing coat*
1 part lime putty
1 part fine washed sand
  This was consolidated with a steel trowel to a smooth surface.

The traditional method requires ample time and freedom from mechanical shocks, both of which are incompatible with the pace of modern building.

A modern lime plaster contains a proportion of quicker-setting material to impart an early strength, so that the time is reduced and the backing is more resistant to mechanical shock. The quick-setting material may be added to the mix or it may be present as an impurity in the lime rendering the lime *hydraulic*. This latter aspect will now be considered in more detail.

Hydraulic lime

The mechanism of the hardening process is usually more complex than that described above, since other materials in the mix may play a part. Moreover, many limes are not derived from the purer varieties of chalk or limestone. Grey chalk and greystone, for ex-ample, contain silica and alumina and during calcination these combine with some of the lime to form a product similar to a Portland cement (*see* Chapter Three). The lime is then called a 'hydraulic lime'. This term arises from its capacity to set and har-den under water. If the hydraulicity is such that the lime will pass a strength test as laid down in B.S. 890 it is classed as 'semi-hydraulic'. The following distinctions are made between the hydraulic and non-hydraulic limes:
(a)   Non-hydraulic lime gradually hardens from the surface in-wards as the atmospheric carbon dioxide penetrates to form calcium carbonate.
(b)   Hydraulic lime hardens by virtue of the chemical reaction and *cross-linking* of the chemical bonds of its hydraulic constituents;

CONSTRUCTION SCIENCE 3

4 CONSTRUCTION SCIENCE 3

this is in addition to the hardening process of non-hydraulic lime.
(c) Mortar prepared from non-hydraulic lime and sand, if kept wet, will not harden. Hydraulic lime, in contrast, will harden under wet conditions by the reaction of its hydraulic content. It is therefore marketed as the dry hydrate, rather than as quicklime which picks up water very quickly, or as putty lime.
(d) In order to obtain maximum yield and ease of working it is necessary to prepare the lime putty some time in advance of its use and to allow it to soak. Whilst a lengthy soaking period can be employed in the case of non-hydraulic lime, such a procedure would destroy any hydraulicity in the lime. Hydraulic hydrates are therefore given a short soak, usually overnight, as a compromise.
(e) Non-hydraulic lime, also called 'fat', 'white', or 'mountain lime', is used with gypsum or cement to increase the workability of these two materials, to enhance the water retention, and to decrease the strength. It is not now used with sand alone.
(f) Semi-hydraulic lime, also called 'grey lime' or 'grey-stone', is used in the same way as non-hydraulic lime and very seldom used with sand alone.
(g) Hydraulic lime, also called 'blue lias', is used in mortar mixed with sand. It must not be mixed with gypsum or cement.

EXPERIMENT 1
(a) To prepare quicklime from limestone or chalk.
(b) To prepare hydrated lime from quicklime.
(c) To examine the setting and hardening process of lime.

Apparatus
Porcelain crucible
Pipeclay triangle
Tripod
Meker burner
Glass rod
Desiccators (2)
Evaporating dish
Shallow tin lids (2)
Ground limestone or chalk
Hydrochloric acid (dilute)
Limewater (a saturated solution of calcium hydroxide)
Lump limestone or marble if solid carbon dioxide is not available ('drikold' is one commercial name for this)

Procedure

A   *To prepare quicklime from limestone or chalk*

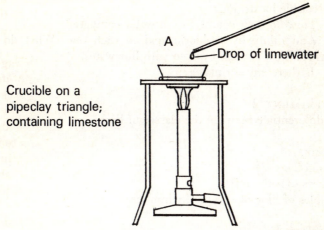

Fig 1   Preparation of quicklime

Set up the apparatus as in Fig 1 and heat strongly. Whilst the calcination is proceeding dip the glass rod in the limewater and hold it at A; incline as shown. Observe the appearance of the drop —a milkiness will confirm the evolution of carbon dioxide; it is due to the formation of insoluble calcium carbonate.

In order that the reaction may be completed in a reasonable time it is necessary to conserve and concentrate the heat. This can be done by placing a metal shield around the crucible.

B   *To prepare hydrated lime from quicklime*
Add water very slowly to a portion of quicklime. Note carefully what happens. Is there any volume change during the slaking?

C   *To examine the setting and hardening process of lime*
Prepare a pad of putty lime in each of the two shallow tin lids. Place them in separate desiccators. Both desiccators should contain calcium chloride to absorb the water which evaporates. One desiccator should contain a piece of solid carbon dioxide or, if this is not available, an evaporating dish containing lump limestone and dilute hydrochloric acid to provide an atmosphere of carbon dioxide in the desiccator. The 'drikold' should be replenished every few days.

After 2 weeks remove the samples and compare their properties:

(a)   Which sample is harder; which has merely set without appreciable hardening?

(b)   How do their solubilities in water compare?

(c)   Add dilute hydrochloric acid to each one. What do you observe? Test any gas evolved with limewater.

(d)   Is there any shrinkage?

EXPERIMENT 2

To differentiate between the classes of lime.

Apparatus

Muslin bags (3)

Beakers (3 × 400 ml)

Samples of lime of varying hydraulicity

Procedure

(a)   Place a few small lumps of one sample of lime in a muslin bag and immerse for a few seconds in clean water. Remove from the water, allow to drain, and empty into a 400 ml beaker.

(b)   Note how long it takes for the onset of slaking.

(c)   If there is no sign of slaking within 2 hours mix the sample into a stiff paste and place a pat of this in a 400 ml beaker. Immerse this in water contained in a 2 litre beaker.

(d)   Check for setting at weekly intervals.

Conclusions

Non-hydraulic lime—slakes very easily.

Semi-hydraulic lime—difficult to slake, compared with non-hydraulic lime; takes up to 2 hours. Sets more slowly under water than the most hydraulic type, which is called:

Eminently hydraulic lime—most difficult to slake. Sets under water within a week.

Selection tests on lime

The British Standard for building limes is B.S. 890. In this standard are found the tests and requirements which apply to limes; if these requirements are met, and the lime is used correctly, then satisfactory results can be expected. Both non-hydraulic and semi-hydraulic limes are covered but not the eminently hydraulic type since it is not used to any significant extent in this country.

The limes are considered under three headings:

Quicklime

Hydrated lime

Lime putty

The last two groups include lime obtained as a by-product from the chemical industry in addition to that from natural sources. Building lime with a high magnesia content is also covered in B.S. 90. These *magnesian limes* are unsuitable for many building purposes but are valuable for such applications as the manufacture of certain types of refractory products used in the building industry.

Tables 1 to 3 list the tests to be carried out and indicate the limits to be applied. They are included as a guide; the British Standard should be consulted if greater detail is required.

It is unlikely that time will be available during the course to carry out all the following tests. Those which are considered to be

Test	High calcium	Semi-hydraulic	Magnesian
Content of carbon dioxide	All less than 6%		
Insoluble matter	All less than 3%		
Content of CaO and MgO	Not less than		
	85%	70%	85%
Content of MgO	Less than 5%		Not less than 5%
Soluble silica	Not less than 6%		
Residue on slaking—			
No 16 sieve	All not greater than 5%		
No 60 sieve	All not greater than 7% total		
Density of standard putty	All not greater than 1·45 g/ml		
Workability of standard putty	All require not less than 14 bumps for a 190 mm spread		
Hydraulic strength	Not less than 0·69 N/mm²		
	Not greater than 2·07 N/mm²		

Table 1   Summary of tests for quicklime (B.S. 890)

Test	High calcium	High calcium (by-product)	Semi-hydraulic	Magnesian
Content of carbon dioxide	All not greater than 6%			
Content of insoluble matter	All not greater than 1%			
Content of CaO and MgO	All not less than 65%			
Content of MgO	Not greater than 4%			Not less than 4%
Soluble salts	Not greater than 0·5%			
Soluble silica			Not greater than 5%	
Fineness—				
No 85 sieve	All not greater than 1% residue			
No 170 sieve	All not greater than 6% total residue			
Soundness	10 mm		10 mm	10 mm
Pat test	Free from pops or pits		Free from pops or pits	
Putty of standard consistence—				
Density	All not greater than 1·5 g/ml			
Workability	All not less than 12 bumps to reach 190 mm spread			
Hydraulic strength (28 days)			Not less than 0·69 N/mm² Not greater than 2·07 N/mm²	

Table 2   Summary of tests for hydrated lime (B.S. 890)

the most instructive are given as experiments; others are only given in outline.

Determination of the carbon dioxide content
The details of this test are not given in B.S. 890. A suitable method is as follows:
    A known weight of the material is heated with dilute phosphoric acid and the gas evolved is dried by passing it through a bubbler containing concentrated sulphuric acid before being absorbed in

Test	High calcium	High calcium (by-product)	Semi-hydraulic	Magnesian
Content of carbon dioxide	All not greater than 6%			
Content of insoluble matter	All not greater than 3%			
Content of CaO and MgO	Not less than 65%		Not less than 60%	Not less than 65%
Content of MgO	Less than 4%			Not less than 4%
Soluble silica			Not greater than 5%	
Soluble salts		Not greater than 0·5%		
Fineness— No 85 sieve	All not greater than 1% residue			
No 170 sieve	All not greater than 6% total residue			
Soundness	10 mm		10 mm	10 mm
Pat test	Free from pops or pits		Free from pops or pits	
Density as supplied	All not greater than 1·5 g/ml			
Separation of water on standing	All not greater than 25 ml			
Putty of standard consistence— Density	All not greater than 1·45 g/ml			
Workability	All not less than 14 bumps to reach 190 mm spread			

Table 3   Summary of tests for putty lime (B.S. 890)

weighed U-tubes containing soda lime. The increase in weight of the U-tubes is the weight of carbon dioxide in the sample.

Determination of the soluble silica content
A known weight of the material is boiled with hydrochloric acid and filtered. The residue is dried, weighed, and heated with hydrofluoric acid and sulphuric acid in a platinum dish, leaving a residue which is weighed. The difference in weight of the two residues is used to calculate the soluble silica content.

Determination of the insoluble matter
The insoluble matter is that which does not dissolve in dilute hydrochloric acid. It is the first residue in the previous determination.

Determination of the residue on slaking
The quicklime is crushed to pass a No 8 sieve with the minimum of fines and then slaked in water kept at boiling point. The product is then boiled for a total of 1 hour and the residue determined by sieving. This procedure is called *isothermal slaking*. The lime putty is reserved for further tests.

EXPERIMENT 3
To determine the residue on slaking of a sample of quicklime.

Apparatus
Slaking vessel (8 litre)
Collecting vessel (buckets totalling 10 l)
B.S. No 8 sieve
B.S. No 16 sieve
B.S. No 60 sieve
Rubber tubing (5 mm diameter, 2 m long)
Oven
Filter cloth
Quicklime

Procedure

A   *Isothermal slaking*
(a)   Boil 4 l of water in the slaking vessel.
(b)   While the water is coming to the boil grind the quicklime to pass through the No 8 sieve with the minimum of fines.
(c)   Add 1 kg of the ground quicklime, in small quantities, to the water, stirring constantly. This should take 5 minutes.
(d)   Boil gently for a total time of 1 hour if the lime is of the high calcium type and for 2 hours if the lime is semi-hydraulic.
(e)   Cool and allow to stand for not less than 12 hours.

B   *Residue on slaking*
(a)   Pass the supernatant liquid and then the remainder through the No 16 sieve superimposed on the No 60 sieve. Collect the liquid passing through the sieves in a bucket.

(b) Wash the residues on the sieves with a moderate jet of water for not more than 30 minutes. Collect the washings in the bucket.

(c) Dry the residues at 100°C in the oven to constant weight.

(d) Calculate the residue on slaking—

on the No 16 sieve,

on the No 60 sieve,

and the cumulative residue on the No 60 sieve.

(e) Pass the milk of lime, including the washings, through the filter cloth and allow to drain. Reserve the lime putty for later tests.

Conclusion

Express the results as percentages of the weight of quicklime taken.

Determination of the workability

The workability is determined on a lime putty corrected to a *standard consistence*. It is the number of *bumps* necessary to spread the putty over an average diameter of 190 mm using the flow table shown in Fig 2.

The flow table consists of a horizontal flat smooth table mounted

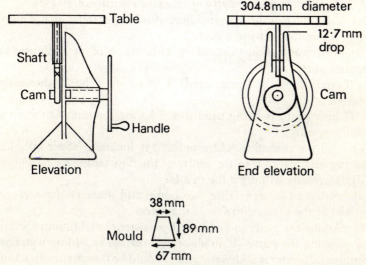

Fig 2   Flow table and mould for workability test (B.S. 890)

on a vertical shaft. The operation of the handle causes the table to be raised and then to fall freely through a height of 12·7 mm. The table is loaded at the rim so that the total weight of the table, rim, and shaft is approximately 6·46 kg. Lime putty placed on the table is subjected to one bump for each revolution of the handle and the workability is the total number of bumps required to attain an average spread of 190 mm in a sample of a standard consistence and size. In order to facilitate the measurement of the spread the table is engraved with rings concentric with its centre having diameters of 80, 110, and 190 mm.

EXPERIMENT 4
To prepare a lime putty of standard consistence and to determine its workability.

Apparatus
Flow table
Conical mould
Trowel
Hydrated lime (quicklime or lime putty may also be used)

Procedure

A   *Preparation of lime putty of standard consistence*
(a)   Mix 500 g of the hydrated lime with 450 g of water for not more than 5 minutes to form a paste.
(b)   Cover and allow to stand for 24 hours. Mix for 3 minutes to form a putty.
   If quicklime is being used it must first be isothermally slaked.
   If lime putty is being used mix 2 kg for 2 minutes to form a uniform putty.
(c)   Fill the conical mould avoiding the inclusion of air bubbles having first placed it in the centre of the flow table.
(d)   Carefully withdraw the mould.
(e)   Subject the material to one bump and measure the average spread of the putty along three diameters.
(f)   Adjust the putty to a standard consistence of 110 mm spread by pressing the putty on an absorbent surface or adding water as required. If water is added the putty should be thoroughly knocked up.

B   *Determination of workability*
(a)   Place the cone of adjusted material in the centre of the table and withdraw the mould.
(b)   Rotate the handle at 1 turn per second. Note the number of bumps to give an average spread of 190 mm.

Conclusion
Record the workability as the number of bumps required to give the putty of standard consistence a spread of 190 mm.

Density of lime putty and the volume yield of lime
The density of lime putty is determined using a density vessel of specified design (Fig 3). The volume yield is calculated from the density of the putty.

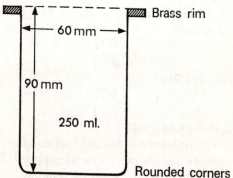
Fig 3   Lime putty density vessel (B.S. 890)

EXPERIMENT 5
To determine the density of lime putty and the volume yield of lime.

Apparatus
Density vessel and cover
Palette knife
Lime putty

Procedure
(a)   Weigh the density vessel and cover.
(b)   Weigh the density vessel full of water and cover.

(c) Fill within 6 mm of the top in 2 increments. Tap the vessel on a protective pad on the bench each time to exclude voids.

(d) Overfill and level off using the palette knife held at an angle of 45°.

(e) Weigh the filled density vessel and cover.

(f) Repeat the settling taps and fill with levelling off if there has been any further settling.

(g) Re-weigh.

(h) Repeat the settling and topping up until constant weight is arrived at.

Conclusion

(a) Density.

$$\text{Density} = \text{Maximum weight/Volume (g/ml)}$$

(b) Volume yield (y).

The volume yield is the volume provided by 1 g of lime.

Putty from quicklime—

$$y = \frac{0 \cdot 57}{d - 1}$$

Putty from hydrated lime—

$$y = \frac{0 \cdot 72}{d - 1}$$

where d is the density in g/ml.

The density determination is required on the putty adjusted to standard consistence and on putty lime as supplied. Putty lime is also tested for the separation of water on standing.

Separation of water on standing

The putty lime is first mixed to give a uniform putty. It is then allowed to stand for 24 hours and the water settling out in this time is measured using a measuring cylinder.

EXPERIMENT 6

To determine the separation of water from a putty lime on standing.

Apparatus

Trowel

Glass cylinder (150 ml diameter, 1200 ml capacity), cover

Measuring cylinder (50 ml)

Lime putty

Procedure
(a)   Mix the lime putty with the trowel to give a uniform putty.
(b)   Weigh 1500 g into the glass cylinder ensuring that no large voids are included. Level off.
(c)   Cover and allow to stand for 24 hours at 20°C.
(d)   Pour off the supernatant water and lime suspension into the 50 ml measuring cylinder and allow to settle for 30 minutes.
(e)   Measure the volume of water.

Conclusion
For the putty to conform to B.S. 890 there should not be more than 25 ml of water settling out.

Fineness
The test for fineness is only applicable to hydrated lime and putty lime. A sample of the lime is shaken with water and poured on to a No 85 sieve superimposed on a No 170 sieve. The residues on the sieves are washed, using a water jet and without separating the sieves, dried, and weighed.

EXPERIMENT 7
To determine the fineness of a sample of lime.

Apparatus
Wide mouth stoppered bottle (approx. 500 ml capacity)
B.S. No 85 sieve
B.S. No 170 sieve
Rubber tubing (5 mm diameter, 2 m long)
Sintered glass filter crucibles (2), grade 2
Buchner filter flask and adaptor
Oven
Hydrated lime or putty lime

Procedure
(a)   Weigh 50 g of hydrated lime (100 g putty lime) into the bottle.
(b)   Shake the stoppered bottle for 30 seconds.
(c)   Pour the contents of the bottle immediately after shaking on to a wet No 85 sieve superimposed on a No 170 sieve.
(d)   Using a moderate jet of water (about 1·22 m head) wash the residue from the bottle on to the sieves, and then wash the sieves for 5 minutes.

(e)   Transfer the residue on the No 85 sieve into a previously dried and weighed filter crucible.

(f)   Dry this residue at 105°C for 1 hour, cool, and weigh.

(g)   Wash the residue on the No 170 sieve for at least 4 minutes but not more than 10 minutes. The washings should now be clear.

(h)   Transfer this residue on to a previously dried and weighed filter crucible.

(i)   Dry this residue at 105°C for 1 hour, cool, and weigh.

Conclusion

Express the residues as percentages of the hydrated lime sample. For lime in the form of putty the water content must be determined by drying 10 g at 105°C.

Soundness

The soundness test is intended to check on the degree of hydration of hydrated limes. Lime which is exposed to too high a temperature during calcination slakes very slowly. If there are particles of overburnt lime in the quicklime it is likely that they will remain unslaked during the manufacture of hydrated lime; manufacturers usually take care to remove these particles after the lime has been hydrated, by taking advantage of the fact that they are coarser than the hydrated particles and are left behind when a blast of air is sent through the product.

An unsound, or incompletely hydrated, lime gives rise to expansion which may be either *local* or *general*.

(a)   Local expansion

This is caused by the presence of nodules of unslaked lime which slowly slake after the work is completed. The surface suffers from defects known as 'blistering', 'blowing', or 'popping', and may become *pitted* as portions fall away.

(b)   General expansion

This may be caused by the presence of finely divided unslaked lime particles distributed throughout the material. They slowly slake as the finish sets causing expansion, not of a local nature this time, but of the finish as a whole. The finish may then *shell* from the backing coat, especially if the key is poor.

EXPERIMENT 8
To test a sample lime for soundness:
(a)   Pat test for local expansion.
(b)   Le Chatelier mould test for general expansion.

A   *Local expansion*

Apparatus
Trowels (2)
Glass plate (460 × 760 mm)
Brass ring moulds (3) (100 mm × 5 mm, thickness 5 mm with an internal taper of 5°)
Base plates (120 mm square), for moulds
Broad palette knife
Ventilated oven
Boiler
Hydrated lime (or putty lime)
Plaster of Paris
Petroleum jelly

Procedure
(a)   Mix 70 g of lime and 70 ml of water on the glass plate.
(b)   Cover and allow to stand for 2 hours.
(c)   Grease the moulds and plates with petroleum jelly.
(d)   Knock up the putty to form a stiff plastic mass adding a little more water if necessary.
(e)   Spread out the putty on the glass plate, sprinkle 10 g of plaster of Paris over the surface, and trowel for 2 minutes.
(f)   Using the palette knife form a flat pat in each ring mould being careful not to include voids. Cut off and smooth the top level using not more than 12 strokes and taking not more than 5 minutes.
(g)   Leave the pats to stand for 30 minutes.
(h)   Put in the oven, controlled at 40°C, for not less than 12 hours.
(i)   Discard and replace any cracked specimens.
(j)   Steam for 3 hours.
(k)   Examine when cool.

B

Conclusion

The lime conforms to B.S. 980 if the specimens are free from pops and pits.

B *General expansion*

Apparatus

Le Chatelier moulds (3) (Fig 4)

100 g weights (3)

Non-porous plates (6) (weight not less than 10 g)

Metal rod (end diameter 17 mm, weight 10 g)

Boiler

Hydrated lime (or putty lime)

Portland cement

Standard Leighton Buzzard sand

Petroleum jelly

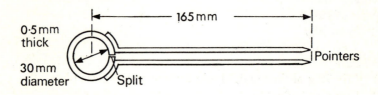

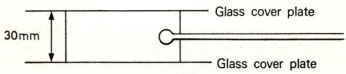

Fig 4   Le Chatelier mould

The Le Chatelier mould is a split cylinder of diameter 30 mm fitted with a pointer on each side of the split. If the material under test expands then the split is widened and the pointers diverge. The difference between the initial and final separation of the pointers gives an indication of the degree of expansion.

Procedure

(a)   Grease the moulds and place them on greased plates.

(b)   Mix 30 g of lime, 10 g of Portland cement, and 120 g of sand.

(c)   Prepare a paste by adding 20 ml of water.
(d)   Fill the moulds, tamping to remove voids.
(e)   Cover with a plate and place a 100 g weight on top of the plate.
(f)   Allow to stand for 1 hour.
(g)   Measure the distance between the pointers.
(h)   Store at 20°C in air of relative humidity 90% for 48 hours.
(i)   Steam for 3 hours.
(j)   Cool and measure the distance between the pointers. Deduct 1 mm to allow for the expansion of the Portland cement. Record the mean value.

Conclusion
For the lime to conform to B.S. 890 the pointer movement should not exceed 10 mm.

Hydraulic strength
The hydraulic strength test is only required for the hydraulic types of lime.

Six 4:1 sand-lime mortar specimens are cast in moulds to provide beams 25·4 mm square by 101·6 mm long. The mortar used is prepared to a standard consistence as determined by the Dropping Ball Test which will be described below. After storage for 28 days in a relative humidity of not less than 90% the beams are immersed in water for 4 to 6 hours and tested immediately. The specimens are rested symmetrically on the sides of two parallel metal rollers 10 mm in diameter and at least 25·4 mm long. Loading is by means of a third roller placed midway between the other two (Fig 5).

EXPERIMENT 9
To prepare a standard 4:1 sand-lime mortar and carry out the hydraulic strength test.

A   *Preparation of the standard 4:1 sand-lime mortar*

Apparatus
Mixer (Fig 6) (A Hobart of capacity approx. 1 litre is suitable)
Dropping ball apparatus and gauge (Fig 7)
Mould (100 mm diameter, 25 mm deep)
Lime
Leighton Buzzard sand

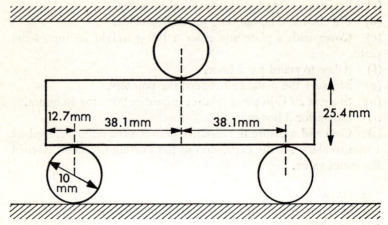

Fig 5   Hydraulic strength test—beam loading (not to scale)

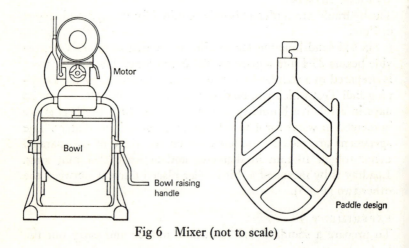

Motor

Bowl

Bowl raising handle

Paddle design

Fig 6   Mixer (not to scale)

Procedure
(a)   Place 500 g of hydrated lime and 2 kg of sand in the bowl of the mixer. Mix dry for 30 seconds. (If putty is being used, then add a weight equivalent to 500 g of dry hydrated lime; similarly for the putty obtained from the isothermal slaking of quicklime.)
(b)   Continue mixing for another 30 seconds while adding the appropriate amount of water to give 10 mm penetration by the dropping ball test (below).

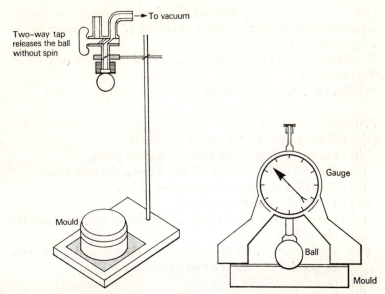

Fig 7   Dropping ball apparatus and gauge

(c)   Mix for a further 60 seconds.

(d)   Clean the paddle and the sides of the bowl, cover with a damp cloth, and leave to stand for 10 minutes.

(e)   Re-mix for 60 seconds and use immediately.

(f)   Dropping ball penetration test.

Fill the mould with the paste mixed as above. This should be done in 10 increments, using the palette knife, to avoid the entrapment of air. Level off.

(g)   Allow the ball to fall 254 mm into the centre of the paste. Note the average penetration from two drops.

B   *The hydraulic strength test*

Apparatus

Six-gang mould (each compartment measuring 101·6 mm long by 25·4 mm square)

Glass plates (2) (235 × 152·4 × 6·4 mm)

Loose fitting hard wood ejectors, wax coated and mounted on a common base

Metal bar (12 mm square, weight 50 g)

Steel tool having a straight edge long enough to span the mould

Sheets of white cotton gauze (2) (222·3 × 136·5 mm)
Extra thick white blotting paper (8 sheets) (22·3 × 136·5 mm)
2 kg weight
Humidity cabinet (an air-tight vessel of volume 0·028 m³)
Test machine
Standard 4:1 sand-lime mortar

Procedure
(a)   Fill the mould by the suction method:
  (i)   Clean the mould and place on a glass plate.
  (ii)   Half fill and ram down 25 times using the metal bar.
  (iii)   Overfill and repeat the ramming down.
  (iv)   Level off using the steel tool.
  (v)   Cover with gauze, 4 sheets of blotting paper, and then the glass plate.
  (vi)   Invert, remove the glass plate from the top, and cover again as in (v). Place the 2 kg weight on top.
  (vii)   Store at 20°C for 2½ hours.
  (viii)   Eject the specimens and leave them uncovered on the ejector in store at 20°C and a relative humidity of 90% for 24 hours from the time of moulding.
(b)   Store the specimens for 28 days over water in the air-tight vessel.
(c)   Immerse in water for 4 to 6 hours.
(d)   Place each specimen in turn in the test machine (*see* Fig 5), and apply the load steadily and uniformly from zero at a rate of not less than 222·4 N per minute and not greater than 889·6 N per minute. Note the breaking load in each case.

Conclusion
The modulus of rupture is calculated in the following manner:
  The bending moment of the beam is given by—

$$\text{B.M.} = \frac{W.l.}{4} = \frac{M.b.d.^2}{6}$$

where W = applied load to break (N).
     l = distance between lower parallel rollers (mm).
    M = modulus of rupture (N/mm²).
     b = breadth of the beam (mm).
     d = depth of the beam (mm).

Questions   Please see appendix at end of book.

Two   Gypsum Plaster

Gypsum plasters are derived from two naturally occurring minerals. These are gypsum, which is calcium sulphate dihydrate ($CaSO_4$ $2H_2O$), and anhydrite, which is anhydrous calcium sulphate ($CaSO_4$). Gypsum is also obtained as a by-product from the chemical industry. Gypsum is first heat treated to remove some or all of the water whilst anhydrite requires only to be pulverised to a convenient size.

When powdered gypsum is heated the proportion of combined water which is driven off is governed by the temperature and duration of the heating. The calcined material is capable of re-combining with water to form a product which is chemically identical to the original gypsum and it does so at a rate determined by the severity of the heat treatment. The rate can also be controlled by suitable additions of accelerators or retarders.

Table 4 is an outline of the process and also shows the five classes of plaster adopted in British Standard 1191.

Setting and hardening

The setting and hardening process of gypsum plaster is quite different from that of lime plaster—and much simpler. As was seen earlier the lime plaster sets and hardens in two distinct stages and this is in addition to the more complex hardening of the hydraulic constituents. A comparison may be made in the following terms:

(a)   The setting and hardening process of gypsum plaster is continuous involving the re-hydration of the calcined material to re-form gypsum.

(b)   The setting and hardening process of lime plaster is a two-stage process—partial drying out followed by carbonation. The mechanism for semi-hydraulic lime plaster is more complex.

23

$$\text{Gypsum } CaSO_4\ 2H_2O$$

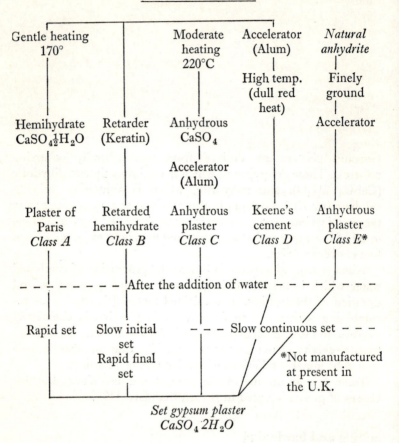

Table 4   The preparation and setting of the five classes of gypsum plaster

(c)   The gypsum is re-formed as a mass of needle-like crystals which interlock to confer considerable mechanical strength. This occurs throughout the plaster and so there is no risk of the coat having a hard skin and soft interior as may happen with lime plaster.

(d)   The setting process is accompanied by a very slight expansion and, once set, properly constituted gypsum plaster is quite free from shrinkage cracks.

(e) Since gypsum plaster undercoats set hard in a matter of hours the finishing coat can be applied without delay.

(f) Since gypsum plaster does not depend on drying for its strength adverse weather conditions do not affect plastering progress.

(g) Gypsum plaster warms up during setting—the re-hydration is an *exothermic* reaction—and so it is not so prone to frost damage as lime plaster.

The point made in (g) above that gypsum plaster warms up during setting can be used to examine the effect of additives on the reaction. In the following experiment the temperature changes are monitored by means of a thermocouple which consists of two wires of different metals joined at each end; one junction is placed in the setting plaster and the other in melting ice. An electromotive force is generated in the instrument the magnitude of which depends on the temperature difference between the two junctions. The e.m.f. can be measured by means of a potentiometer, as below; or the current which flows can be measured by means of a galvanometer. A more convenient method is to use a pen recorder, if one is available, which produces a temperature/time graph automatically.

EXPERIMENT 10
To examine the effect of additives on the setting of gypsum plaster (plaster of Paris).

Apparatus
Potentiometer and tapping key, switch
Accumulator (2 volt)
Galvanometer
Resistance box (10000 ohms), or rheostat
Thermocouple
Plastic cups (drinking cups supplied by 'Mono-containers' are suitable)
Beakers (100 ml)
Vacuum flasks (2)
Stopclock
Burette (50 ml)
Plaster of Paris
Solutions: (a)   Colloids—
               1% gelatine
               1% glue size

    (b)   Crystalline salts—
          1% sodium chloride
          1% alum
          10% alum

Procedure

(a)   Set up the apparatus as in the diagram: the cold junction is ice in a vacuum flask; the hot junction is boiling water in a beaker.

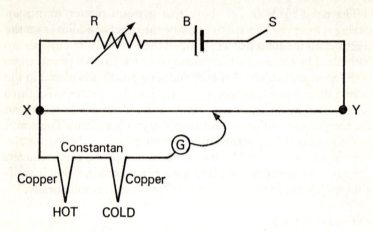

(b)   Vary the resistance R to obtain a balance point D near to the end of the potentiometer wire XY.

(c)   Place the hot junction in water at room temperature and obtain a new balance point.

(d)   Place 50 g of plaster of Paris in a plastic cup positioned on corks inside a vacuum flask.

(e)   Run 30 ml of water into the plaster and stir carefully with the hot junction.

(f)   Obtain a balance point every 30 seconds.

(g)   Construct a graph plotting the potentiometer reading against the time.

(h)   Repeat (d) to (g) using each of the solutions in turn.

Conclusion

Comment upon the effect of the solutions on the setting of gypsum plaster.

The classes of gypsum plaster and their uses

The following quotation is taken from British Standard Code of Practice CP 211:1966.

(1) *Mixes based on gypsum plasters*

Plastering mixes based on gypsum plasters show wide variations in properties and require differing treatments for their effective use. Materials covered by British Standards may be divided into two main groups, those based on calcium sulphate hemihydrate (plaster of Paris and retarded hemihydrate gypsum plaster) and those based on anhydrous calcium sulphate (anhydrous gypsum plasters and Keene's).

The hemihydrate plasters set quickly, the start of the setting process being delayed in the retarded plasters. Plasters based on anhydrous calcium sulphate, on the other hand, have a comparatively slow continuous setting process and in practice should not be allowed to dry before the setting process is complete.

The setting of all these plasters is accompanied by an expansion which is variable in amount, but when the setting process is complete little further movement takes place during drying. For this reason it is unnecessary to ensure that thorough drying of one coat has taken place before the application of the following coat (condition essential for mixes containing cement), but sufficient suction and strength should have been developed to enable good adhesion to be obtained.

No gypsum plaster should be permitted to remain under persistently damp conditions after it has set as this causes weakening and disintegration.

(2) *Mixes based on lime, gauged with a gypsum plaster*

Plastering mixes of gypsum plaster, sand and lime, are easily worked and have a working time which varies with the type of gypsum plaster used. Gypsum plasters expand on setting and tend to restrain the drying shrinkage of the lime. Sufficient time should be permitted for one coat to dry so that it has adequate suction and strength to receive the next one. The time intervals necessary vary widely.

Mixes of lime and gypsum plaster are used for final coats when a fairly soft finish is suitable. The hardness and strength of the finish increase as the proportion of gauging material increases. Care should be taken to ensure that the strength of the finish is not incompatible with the nature of the backing material.

Gypsum plaster is commonly mixed with sand when applied as an undercoat. For this purpose undercoat or dual purpose plaster should be used. Lime and gypsum plaster are often used in the same mix, with or without sand, since, as was mentioned above, the presence of lime confers improved workability and reduces the strength. It also quickens the set.

Some of the more important points to be borne in mind when mixing are:

(a)   Do not mix different classes of plaster or different brands of the same class.

(b)   Careful attention must be paid to the sand quality; the presence of loam or clay will affect the setting time.

(c)   Too much sand in an undercoat reduces the strength; more sand is required for undercoats on brickwork than for those on plasterboard or concrete.

(d)   Never mix cement and gypsum; chemical action may cause disintegration.

(e)   Use clean water; drinking water is generally the most suitable.

(f)   Lime should only be used with Classes A and B plasters and should be of the non-hydraulic or semi-hydraulic variety. Magnesian lime should not be used in the same mix as gypsum plaster.

(g)   Gypsum mixes cannot be 'knocked up' again.

Class A. Plaster of Paris

Plaster of Paris is produced by gently heating crushed gypsum. It is not used alone for plastering owing to its rapid set but is valuable for ornamental mouldings and for repair work.

Class B. Retarded hemihydrate gypsum plaster

Retarded hemihydrated plaster, or *hardwall* plaster, is chemically the same as plaster of Paris but to render it suitable for general plastering the rate of crystal growth is slowed down by the incorporation of retarders. The manufacturers' directions must be carefully followed; the time recommended for mixing and application must be adhered to if a satisfactory set is to be obtained, since it is manufactured in various grades to suit particular situations:

Browning plaster
Designed for use, when gauged with sand, as a floating coat on
brick walls, concrete bricks, clay partition blocks, clinker blocks,
no-fines concrete, stone masonry, etc.

It should never be used neat since the setting time is too slow;
when sand is added in the proportion plaster:sand = 1:3 the
setting time is of the order of two to three hours.

Floating coats on gypsum boards and polystyrene slabs are
formulated using a Browning plaster to which nylon fibre has been
added. This is called 'haired plaster', as in the days when animal
hair was used. It is generally used with $1\frac{1}{2}$ parts of sand.

Metal lathing plaster
Contains rust inhibitors and should be used gauged with sand as a
floating coat where metal contact is likely.

Finish plaster
The setting time is 2 to $2\frac{1}{2}$ hours when used neat but somewhat
shorter when mixed with lime.

Board finish plaster
The setting time is approximately 1 hour. It should be used neat
since the addition of lime would cause the finish to shrink and
thereby lose adhesion to the board.

Class C. Anhydrous gypsum plaster
Anhydrous gypsum plaster is produced by heating gypsum at
about 220°C thereby removing almost all of the chemically combined water. During manufacture accelerators are added to give
a workable setting time.

A high standard of finish is possible by virtue of the slow, continuous set and the plaster is valuable where durable, hygienic,
and easily decorated surfaces are required. It is not suitable for
direct application to plasterboard, fibreboards, or similar wallboards.

When used in a floating coat on cement and sand, or cement,
lime, and sand undercoats it reduces the risk of the shrinkage
cracking being transmitted through to the final coat. In plain
finishes Class C plaster is used neat.

Since the gradual set allows time for the plaster to be worked,
a variety of textured finishes can be obtained using a range of tools;

a few of those suggested are wood float, sponge, hair or rubber brush, felt float, and sacking tied round the trowel or float. The plaster may be used neat, at a stiffer consistency than usual, or in a 1:1 mixture (by volume) with washed sand, depending on the desired effect.

Class D. Keene's cement
Keene's cement is produced by heating gypsum to a temperature sufficient to drive off all the water. Accelerators are added during manufacture to reduce the setting time which otherwise would be very long indeed.

It is hard and tough and can be worked to a perfectly true surface. It is always used neat and applied as a finishing coat 3 mm thick to a hard undercoat, preferably composed of Portland cement and sand, which has been scored to form a key and allowed to dry out thoroughly. It must not be applied direct to plasterboard, fibreboard, or similar wallboards.

Paint adhesion to this class of plaster can be very poor. Under exceptionally fast drying conditions it is advisable to *follow the trowel* with primer. It should be applied as soon as the plaster is sufficiently hard not to be scored by the brush and should be *sharp*, containing only the minimum of oil necessary for pigment dispersion.

Selection tests on gypsum plasters

The British Standard for gypsum building plasters is B.S. 1191 which has two parts; part 1 deals with dense types while part 2 is devoted to lightweight types. In this section consideration will be given to the former and the tests given in detail are those which may reasonably be carried out in the building laboratory. For greater details of the other test reference should be made to B.S. 1191.

Table 5 gives the tests to be applied to each class of plaster at present manufactured in this country.

Where the tests are given as experiments they are based on B.S. 1191. If time and conditions do not allow them to be followed exactly they can be modified, for example, by changing the specimen ageing, without detracting from their value to the course.

	Class A	Class B	Class C	Class D
Chemical composition	√	√	√	√
Freedom from coarse particles	Max. quantity retained on a B.S. 14 sieve			
	5%	1%	1%	1%
Soundness	√	√	√	√
Transverse strength	Not less than			
	24·5kN/mm²	12·2kN/mm² (sanded)	Not applicable	
Mechanical resistance	Not applicable	Maximum penetration		
		5 mm	4·5 mm	4 mm
Expansion on setting	Not applicable	Max. 0·2%	Not applicable	

Table 5   Tests required on gypsum plasters

Class A plaster

Freedom from coarse particles
The plaster is shaken on a B.S. 14 sieve for 5 minutes when not
more than 5% should be retained. Lumps should be broken down
with the fingers and not by rubbing on the sieve.

Soundness
The plaster is gauged with water to produce a smooth cream which
is placed in 6 moulds each of diameter 101·6 mm and depth 6·3
mm. The moulds are then placed in a damp closet having an
atmosphere of relative humidity 80% for a period of 16 to 24
hours after which they are steamed for 3 hours.

The pats should show no sign of disintegration, pitting, or
popping.

Transverse strength
The plaster of Paris is first stabilised by exposure to air for 3 to 4
days in a layer not greater than 12·7 mm deep. A continuous
stream of air should be passed over the surface. The air humidity
should be 65% and the temperature 20°C. If a controlled humidity
cabinet is not available this humidity can be achieved by placing a
wide dish containing a saturated solution of ammonium nitrate
adjacent to the plaster.

After stabilisation a plaster of standard pourable consistence is made up, cast into moulds, aged, and dried. The transverse strength is measured on six such specimens.

EXPERIMENT 11

To measure the transverse strength of a Class A gypsum plaster.

Apparatus
Humidity cabinet
Oven
Consistence mould (a hollow, corrosion-resistant metal cylinder of diameter 30 mm and height 50 mm with the ends square to the longitudinal axis; it is centred on a metal base plate)
Transverse strength moulds (6-gang mould, each section measuring $101 \cdot 6 \times 25 \cdot 4 \times 25 \cdot 4$ mm, as was used for lime plaster in Experiment 9)
Transverse test machine
Open tray ($30 \times 30 \times 24 \cdot 4$ mm approx.)
Plaster of Paris
Sodium citrate

Procedure
Stabilise the dry plaster.

A   *Preparation of the standard pourable consistence*
(a)   Sprinkle 100 g of plaster over a known volume of water containing 0·1 g of sodium citrate. This should take 30 seconds.
(b)   Allow to stand for 30 seconds.
(c)   Stir for 60 seconds.
(d)   Fill the cylinder and level off.
(e)   Allow 30 seconds to elapse *after mixing* and lift the cylinder vertically so that the mix spreads over the base plate.
(f)   Measure the maximum and minimum diameters of the spread and record the mean.
(g)   Repeat, varying the plaster:water ratio until a mean spread of 78 to 80 mm is achieved.
(h)   Record the plaster:water ratio finally arrived at.

B   *Preparation of the test specimens.*
(a)   Sprinkle 450 g of plaster over the appropriate amount of water required to produce the standard consistence taking 30 seconds to do so.

(b)   Allow to stand for 30 seconds.
(c)   Stir for 60 seconds.
(d)   Fill the 6 transverse strength moulds ensuring that no voids are included.
(e)   Allow to stand for 24 hours in the humidity cabinet set at 90% R.H. and 20°C.
(f)   Scrape the specimens level with the top of the moulds and dry at 35 to 40°C to constant weight.

C   *Testing the specimens*
(a)   Place the specimens symmetrically in turn on the rollers (12·7 mm diameter) at 76·2 mm centres.
(b)   Apply the load on a third central roller at not less than 49·03 N per minute and not more than 88·90 N per minute and note the breaking load.

Conclusion
Calculate the modulus of rupture for each specimen.

Class B plaster

All Class B plasters must conform to B.S. 1191 with respect to chemical composition and soundness. In addition the following tests must be applied to the different types within this class.

Type a.1. Browning plaster
Transverse strength.

Type a.2. Lathing plaster
Free lime must be not less than 3% by weight on despatch. Transverse strength.

Type b.1. Finish plaster
Coarse particles
Mechanical resistance.

Type b.2. Board finish plaster
Coarse particles.
Mechanical resistance.
Expansion on setting to be not greater than 0·2% in 1 day.

Freedom from coarse particles
Not more than 1% should be retained on the B.S. 14 sieve.

Soundness
The plaster is gauged with sufficient water to produce a stiff
plastic paste and then tested in the same manner as a Class A
plaster.

Transverse strength
This test is only carried out on the undercoat types. The plaster is
first stabilised in the manner employed for Class A plaster after
which a paste of standard undercoat consistence is made up and
checked using a dropping ball penetrometer (Fig 7). The paste in
this case has the composition 1 part plaster to 3 parts sand (by
weight).

Test prisms measuring $101 \cdot 6 \times 25 \cdot 4 \times 25 \cdot 4$ mm are prepared and
tested.

EXPERIMENT 12
To measure the transverse strength of a Class B plaster.

Apparatus
Spatula (stiff)
Spatula (flexible)
Humidity cabinet
Oven
Dropping ball penetrometer and ring mould
Open tray
Transverse strength moulds (6-gang)
Brass rod (6·35 mm square section)
Transverse test machine
Browning or lathing plaster
Standard Leighton Buzzard sand

Procedure
Stabilise the plaster (as in Experiment 11).

A   *Preparation of the standard undercoat consistence*
(a)   Dry mix 200 g of plaster and 600 g of sand.
(b)   Add the dry mix to water. This should take 30 seconds.
(c)   Mix for 1 minute with the stiff spatula.

(d)   Test the mix for consistence:
 (i)   Fill the ring mould in ten increments using a flexible spatula
       to eliminate voids.
 (ii) Drop the 25·5 mm diameter methylmethacrylate ball of
       weight 10 g from a height of 254 mm.
 (iii) Measure the penetration.
(e)   Repeat (d) with mixes containing different proportions of
water until a penetration of 9 to 10 mm is obtained. This paste has
the standard consistence.

B   *Preparation of the test specimens*
(a)   Half fill the 6-gang mould and consolidate by tamping 10
times with the brass rod along the length of the mould.
(b)   Fill the mould and similarly consolidate the top layer.
(c)   Leave for 24 hours under a damp cloth or in a humidity
cabinet.
(d)   Scrape the plaster level with the top of the mould.
(e)   Remove the prisms from the mould and dry them to constant
weight in a well-ventilated oven at 35 to 40°C.

C   *Testing the specimens*
(a)   Place the specimens symmetrically in turn on the rollers as
in Experiment 11.
(b)   Apply the load on a third central roller at not less than 49·03
N per minute and not more than 88·9 N per minute and note the
breaking load.

Conclusion
State whether the plaster conforms to B.S. 1191 undercoat plaster.

Mechanical resistance
This test is only carried out on final coat plaster (*see* later for
Classes C and D).
   The plaster is first stabilised in the usual manner and then made
up into a paste of standard final coat consistence. A steel ball is
dropped on to a moulded, aged specimen and the indentation
caused taken as a measure of the mechanical resistance.

EXPERIMENT 13
To measure the mechanical resistance of a Class B plaster by the
dropping ball test.

Apparatus
Spatula (stiff)
Spatula (flexible)
Humidity cabinet
Oven
Open tray
Brass rod (6·35 mm square section)
Dropping ball penetrometer (methylmethacrylate ball)
Ring mould
Retort stand and clamp
Mould (6-gang, each section 101·6 × 25·4 × 25·4 mm)
Steel ball (diameter 12·7 mm, weight 8·33 g)
Straight tube (internal diameter 15·8 mm, length 1·72 m)
Class B finish plaster

Procedure
Stabilise the plaster (as in Experiment 11).

A   *Preparation of the standard final coat consistence*
(a)   Add 200 g of plaster over 30 seconds to a known weight of water.
(b)   Allow to soak for 1 minute and then mix with a stiff bladed spatula for 2 minutes.
(c)   Test the mix for consistence.
  (i)   Fill the ring mould in ten increments using a flexible spatula to eliminate voids.
 (ii)   Drop the 25·4 mm diameter methylmethacrylate ball from a height of 254 mm.
(iii)   Measure the penetration.
(d)   Repeat (c) with mixes containing different proportions of water until a penetration of 15 to 16 mm is obtained. This paste has the standard consistence.

B   *Preparation of the test specimen*
(a)   Add 400 g of the plaster over 30 seconds to the appropriate amount of water.
(b)   Allow to soak for 1 minute and then mix with a stiff-bladed spatula for 2 minutes.
(c)   Fill the gang mould in 2 layers tamping each layer 10 times with the brass rod. Level off.
(d)   Leave for 24 hours under a damp cloth or in the humidity cabinet at 90% R.H. and a temperature of 20°C.

(e)   Remove the specimens from the mould and dry to constant weight in the oven at 35 to 40°C with good ventilation.

C   *Testing the specimens*
(a)   Place a smooth face of the specimen being tested upwards on a smooth firm surface. The specimen should have smooth faces where it has been in contact with the mould.
(b)   Clean and polish the steel ball and release it from rest down the tube which is supported firmly in a vertical position with its lower end 101·6 mm above the top surface of the specimen.
(c)   Obtain 8 impressions—one on each of the opposite face of 4 rods avoiding air bubbles and blemishes. Arrange that the impressions are not more than 6·3 mm from the centre line and not within 12·7 mm of the ends of the rod.
   Measure two *diameters* at right angles for each impression.
(d)   Record the mean of the 16 measurements.

Conclusion
State whether the plaster conforms to B.S. 1191 for a Class B finish plaster.

Expansion on setting
This test is only carried out on Class B board finish plaster.
   Measurement is made of the linear expansion on setting in continuous damp air storage of neat plaster gauged to standard final coat consistence.

EXPERIMENT 14
To measure the expansion on setting of a Class B plaster.

Apparatus
Humidity cabinet
Open tray
Spatula (stiff)
Extensometer (Fig 8)
Class B board finish plaster
Petroleum jelly
Thin, non-absorbent, shiny paper (100 × 100 mm approx.)

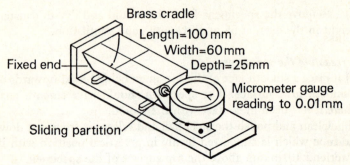

Brass cradle
Length=100 mm
Width=60 mm
Depth=25 mm
Fixed end
Micrometer gauge
reading to 0.01 mm
Sliding partition

Fig 8   Extensometer used for measuring the expansion on setting
of plaster

Procedure
(a)   Stabilise the plaster (as in Experiment 11).
(b)   Add 200 g of plaster over 30 seconds to the appropriate amount of water to produce a plaster of standard final coat consistence. (The amount of water to use is first determined as in Experiment 13.)
(c)   Allow to soak for 1 minute and then mix with a stiff-bladed spatula for 2 minutes.
(d)   Grease the cradle of the extensometer and line it with a thin non-absorbent shiny paper.
(e)   Fill the cradle with the plaster and adjust the zero point as follows:
  (i)   Move the movable partition very slightly forward, clear of the end, to eliminate backlash.
  (ii) Bring the plaster sound against the movable partition.
  (iii) Make any necessary zero adjustment on the dial.
(f)   Place the extensometer in the humidity cabinet set at 90% R.H. and a temperature of 20°C. Note the zero and leave for 24 hours.
(g)   Take the final reading.

Conclusion
Calculate the percentage linear expansion

$$= \frac{\text{Difference in dial reading (in mm} \times 10^{-2})}{100}$$

State whether the plaster conforms to B.S. 1191 for a Class B Board finish plaster.

Class C and Class D plaster

Freedom from coarse particles
Not more than 1% should be retained on the B.S. 14 sieve.

Soundness
The plaster is gauged with sufficient water to produce a stiff plastic paste which is placed in 6 moulds each of diameter 101·6 mm and depth 6·3 mm. The moulds are then placed in a damp closet having an atmosphere of relative humidity 80% for a period of 3 days after which they are steamed for 3 hours.

The pats should show no sign of disintegration, pitting, or popping.

This procedure only differs from that for Class B plaster in the period of humid ageing; the stipulation for Class B plaster is 16 to 24 hours, whereas Class C and Class D plasters require 3 days.

Mechanical resistance
The mechanical resistance test is carried out in the same manner as for Class B plaster (p. 35) except that a different mixing time is stipulated after soaking.

The plaster is added, over 30 seconds, to the appropriate amount of water to give a paste of standard final coat consistence and allowed to soak for 1 minute. It is then mixed with a stiff bladed spatula for:
30 seconds—Class C
60 seconds—Class D
The test then follows the same procedure as for Class B plaster and the plaster must not have a penetration diameter greater than:
4·5 mm—Class C
4·0 mm—Class D

Possible defects in plasterwork and some suggested remedies

Adhesion loss
If a complete gypsum job fails to adhere the cause is generally the use of a strong finishing coat on a weak backing. The backing coat must not contain an excessively loamy sand nor must the sand

content exceed the proportion recommended for the particular
background to which it is applied.

Table 6 gives recommended formulations for various backgrounds.

Background	Total No. of coats	B.S. 1191 classification	Plaster : sand (by volume)	Thickness (mm) U/c	Finish
Clay brick	2	Browning B–a.1	1:3	11	2
*Concrete brick, clinker brick	2	Browning B–a.1	1:2	11	2
*Engineering bricks	2	Browning B–a.1	1:1½	11	2
Wood wool slabs	2	Metal lathing B–a.2	1:2	11	2
Expanded polystyrene	2	Haired B–a.1	1:1	11	2
Expanded metal lathing	3	Metal lathing B–a.2	render 1:1½ float 1:2	11***	2
Solid gypsum partition (50·8 mm)	3	Haired B–a.1	render 1:1½ float 1:1½	15****	2
**Gypsum baseboard, lath, and dry partition	2	Haired B–a.1	1:1½	8	2
**Gypsum baseboard, lath, and dry partition	1	Board finish B–b.2	neat	—	2

*raked joints	***from face of lath
**grey side	****each side

Table 6   Uses of plasters over normal backgrounds
(Gypsum Plasterboard Development Association, 1968)

Sand and lime, or sand, cement, and lime backings are more
prone to this defect where it can be traced to either the application
of the finish before the backing is thoroughly dry or the lack of a
proper key owing to inadequate cross scratching.

When gypsum plaster is applied to a porous surface some of the
plaster solution is drawn into the pores by suction or capillary

action. As the plaster sets crystals form within the pores and it is the cross-linking of these crystals with those of the main body of the plaster which provides the bond. The degree of porosity of the substrate can be understood to have a major influence on the strength of the bond. If this is low, examples are glazed tiles and painted surfaces, the adhesion will be poor. On the other hand if it is very high, as it is with some lime-based concretes, the rapid absorption of the water into the background will result in a deficiency of plaster solution at the surface with a consequent reduced growth of the bond forming crystals.

The use of a plaster/vermiculite backing coat has been found effective in minimising the effect of surface porosity on bond strength. The vermiculite retains the water and reduces the rapid loss of water into the background.

Adhesion is improved by increasing the effective area of plaster actually coming into contact with the background. This can be achieved by dampening the surface a few minutes before the plaster is applied to displace air from the surface pores. Removal of air bubbles mechanically entrained in the plaster during mixing by applying the first layer of backing coat as a very tight thin lay also increases the area of contact.

For adequate adhesion, therefore, it is most important that the background must be suitably pre-treated, the plaster must be correctly formulated, and the application instructions carefully followed.

Assuming that these conditions have been satisfied and adhesion loss still occurs it may be due to movement of the background relative to the plaster which may arise from the following:
(a)   Drying shrinkage of the background after plastering.
(b)   Extreme thermal movement between the background and the plaster as may occur with some types of lime-based concrete.

If a weak backing coat is the cause of the defect the only remedy is to remove all the plaster and renew it. In other cases all loose plaster should be removed, the backing coat should be made good, allowed to dry thoroughly, and roughened, and the final coat applied. It is recommended that concrete surfaces be treated with a PVA emulsion.

Blowing or pitting
This is a defect in which a conical piece of the finished plastered surface is blown out. It may be due to moisture reaching an

imperfectly slaked nodule of quicklime in the work or to the presence of stale cement or unset cement arising from poor mixing.

Cracking and crazing
Cracks appearing in the finished work can be considered under three headings:

(a)   *Fine hair cracks*
These are due to plaster shrinkage and most commonly occur in work with a sand and lime, or a sand, lime, and cement floating coat but can also occur in a gypsum plaster containing a very loamy sand or a considerable excess of lime in the finishing coat. They are generally a sign that the finishing coat has been applied before the floating coat is thoroughly dry and its shrinkage has been completed.

Excessive use of the water brush and trowel on the final coat can *kill* the gypsum causing crazing of the surface with fine hair cracks.

Treatment is extremely difficult. The usual procedure is to paper the surface instead of using paint.

(b)   *Settlement cracks*
These are more clearly defined than hair cracks and more directional. As their description implies they are caused by movement of the structure of the building. In ceilings movement may be a consequence of shrinkage or warping of timber joints, shrinkage of *in-situ* concrete, or thermal and/or structural movement of precast units. On walls they are generally found around concrete lintels and window sills where they run diagonally.

When settlement is considered reasonably complete the cracks should be cut back and flush filled with a proprietary filler.

(c)   *Dry-out*
Should the gypsum plaster dry before it has set it will stay soft and powdery and break into a large number of very fine cracks. This is most likely to happen to the slower setting grades of finish in warm dry weather or if central heating is turned on too early, especially if the coat is too thin.

Decoration faults
The principal factors which influence the painting of any plaster surface are:

The water content of the plaster.
The porosity of the plaster surface.
The presence of efflorescent salts.
Alkalinity.

Decoration faults may arise from neglect of the consideration of these factors and, for convenience, we shall consider them under three headings:

(a) *Sealing of water*

Paint failure on gypsum plaster is almost invariably due to decorating before the work is completely dry. The decorated surface takes on a patchy appearance, blisters, and in some cases the whole film of paint may part from the plaster.

In situations where it is not possible to delay decoration until all the plaster has fully dried the paint finish should be permeable to water vapour. Matt finish emulsion paints, vinyl water paints, and porous wallpaper allow the plaster to breathe.

Not only are these paints useful as a first decoration but they also act as sealers, evening out variations in plaster porosity. If the variations are particularly great a solvent-based priming coat should be applied after the plaster has dried out but before the full emulsion paint system is applied.

Decoration faults arising from the sealing of water are remedied by redecorating when the work is completely dry.

(b) *Efflorescence*

Efflorescence is the formation of a crystalline deposit on the surface. If salts such as the sulphates and carbonates of sodium, potassium, magnesium, and calcium are present in the plastering materials they may be carried by the mixing water to the surface. As the water evaporates they are left behind as crystals either in the form of a white fluffy deposit or as a tight, highly porous, translucent film. Efflorescence may also arise from the use of *salty* bricks.

The usual outcome of efflorescence is the flaking of the paint film from the plastered surface. There are a number of courses of action which may be taken to control it. One is to remove it by brushing down repeatedly until it has ceased to appear. Another is to apply an *efflorescence barrier*, such as a specially formulated oil or varnish, over the surface. If the bricks are at fault a bituminous sealer used prior to plastering has been found to be efficacious. If this procedure is adopted a gypsum plaster must be used since cement gauged plasters do not bond well on a bituminous surface.

(c) Chemical attack

If an oil paint is applied to a lime-containing plaster which is still new and wet saponification of the oil may occur resulting in blisters and yellow oily runs in extreme cases. The paint pigments may also be attacked and discoloured. The defect will not occur with a complete gypsum finish.

Mould growth

Under certain constant conditions of temperature and humidity microscopic mould spores which permeate the atmosphere are liable to take root in small particles of organic dust. The flowering heads of these micro-organisms are recognised as *mould growth*.

Its occurrence is a sign of dampness either in the plaster or in the atmosphere and is best prevented by maintaining good ventilation and by turning on the heating as soon as possible. Damp cold conditions must be avoided.

The mould growth should first be removed by brushing with a soft brush under conditions of good ventilation. The affected areas are then treated with a solution of a fungicide such as 'Topane WS' (I.C.I.) according to the manufacturers' instructions. This treatment will kill all the mould and any remaining spores which have not germinated.

Rust staining

Rust stains can develop in plaster applied over conduits and around switch boxes, etc. Whilst it is moisture which is responsible for corrosion and not gypsum the latter, being slightly acidic, accentuates the attack. Lime, on the other hand, is an inhibitor.

The affected area should be cut out, the metal given a coat of aluminium paint, and the work made good with gypsum metal lathing plaster which contains a rust inhibitor. An alternative remedy is to cover the rust spot with a coat of 'Kemobel' (I.C.I.) or similar paint followed by an undercoat such as 'Dulux' undercoat (I.C.I.). This will prevent the rust stain being drawn through to the final decoration.

Setting difficulties
It sometimes happens that the plaster has a setting time longer or shorter than expected.

(a) *Long setting time*
When a long setting time is experienced with sanded plaster it is usually attributed to the use of unsuitable sand.

(b) *Short setting time*
A short setting time will be experienced with plaster fresh from the works which has not been allowed to cool before use, plaster gauged with dirty water, mixed with partially set plaster from a previous gauging, stored under damp conditions, or kept too long. Some sands also speed up the setting time.

Shelling
As the undercoat or background shrinks a gypsum plaster may *shell* from a cement gauged undercoat which has not been adequately scored to provide a key. If Class C or D plasters shell this may be due to delayed hydration resulting in expansion. Heavy condensation or decoration before the plaster has fully hydrated are the common causes. Class B plaster is quicker setting and so is not prone to delayed hydration. If the undercoat is highly absorbent it could happen that the retarder is concentrated at the interface. In such cases the use of a P.V.A. bonding agent is recommended before remedial re-skinning.

Questions Please see appendix at end of book.

Three Cement

A cement is a material used for bonding together other materials. We shall see in the chapter on building stone that cementitious materials occur in nature and are responsible for holding together the particles in many rocks where they play a major part in determining the properties of the rocks. The cements of particular interest in building originated in the Portland cement, invented in 1824, which was so named owing to its similarity, on setting, to Portland stone. It was manufactured by heating a mixture of clay and limestone.

Manufacture

Portland cement is now manufactured by mixing chalk, or limestone, with clay or shale and sintering it at approximately 1500°C in a rotary kiln to produce clinker. In order to ensure a uniform product it is essential that the raw material fed to the kiln is an intimate mixture and it is for this reason that in many cement works the *wet process* is used. Here the clay and chalk or limestone is ball-milled in the presence of water to produce a slurry before being passed to the kiln. The kiln is a long steel cylinder lined with refractory bricks. It is rotated about its own axis, which is inclined slightly to the horizontal, and fired by pulverised coal or fuel oil blown in through the lower end. The slurry is fed continuously into the upper end and, as it moves to the lower end, it enters progressively hotter zones. In the fore-part of the kiln water is driven off and then the calcium carbonate is decomposed into calcium oxide (quicklime) and carbon dioxide. When the material reaches the hotter zones complex chemical reactions take place between the alumina, silica, and lime to form *clinker*. The clinker is then cooled, mixed with gypsum, and ball-milled to a fine grey powder.

The addition of a carefully controlled proportion of gypsum (4 to 7%) is necessary to prevent what is known as 'flash setting'. It reacts with the small amount of tricalcium aluminium hydrate formed during the sintering which, if it were not removed in this way, would react with the water added to give an undesirably rapid stiffening of the paste.

Types of Portland cement

Ordinary Portland cement (B.S. 12:1958)
The cement manufactured by the above process is generally referred to as 'ordinary Portland cement' (O.P.C.). It is the most commonly used cement having a medium setting time and is suitable for all normal purposes.

There are also modified forms of Portland cement having slightly different properties such as rate of gain of strength and resistance to chemical attack. It should be noted, however, that there is little difference in the ultimate strength of the various types.

In addition to the modified Portland cements there are also special cements formulated by mixing the true Portland cement with other cementitious materials.

Rapid-hardening Portland cement (B.S. 12:1958)
Rapid-hardening Portland cement is similar chemically to O.P.C. but is finer ground and the proportions of its constituents are slightly different. It is used where an early strength is required to allow construction to proceed more rapidly or where formwork cannot be left in position for long. It must be realised that this cement is not a quick-setting cement—its specified setting time is the same as that of O.P.C.—but rather that it achieves its ultimate strength more rapidly.

Extra-rapid-hardening Portland cement
This cement is manufactured by grinding calcium chloride (1 to 2%) into rapid-hardening Portland cement. Alternatively the calcium chloride can be added to the water immediately before mixing the concrete. It is valuable for urgent repair work and for cold weather concreting (2 to 5°C). This latter feature arises from the increased rate of heat evolution accompanying the increased rate of hydration. The addition of the accelerator also decreases the

setting time and so this cement must be placed quickly. The resistance to sulphate attack is reduced and the drying shrinkage is increased. On the other hand the resistance to erosion and abrasion is improved.

Low-heat Portland cement (B.S. 1370 :1958)
Having a slightly lower proportion of the more rapidly hydrating compounds than O.P.C. this cement hardens more slowly. The heat evolved after 28 days is only some 70% of that of O.P.C.

Its use in the construction of large concrete masses reduces the risk of cracking.

Sulphate-resisting Portland cement (B.S. 12:1958)
The composition of this Portland cement has been adjusted to resist attack by sulphates (especially magnesium and sodium sulphates) which are present in sea-water, some ground waters, and some building materials. They cause the concrete to increase in volume and finally to disintegrate.

The use of this cement is particularly important in marine structures between the high and low water marks since the attack is known to be assisted by alternate wetting and drying.

Blast-furnace Portland cement (B.S. 146:1958)
Blast-furnace Portland cement is manufactured by intergrinding O.P.C. with up to 65% blast-furnace slag. It has a lower heat of hydration than O.P.C., is more resistant to chemical attack, and reaches an ultimate strength similar to that of O.P.C. but not so quickly.

Masonry cement
Masonry cement is made by grinding O.P.C. with limestone, an air entraining agent, and a plasticiser.

The cement has a lower strength than an O.P.C.-sand mix and hardens more slowly; both qualities are advantages in a mortar. It also confers other benefits; the mortar is more plastic, water retentive, and less liable to shrink.

White and coloured Portland cements
White Portland cement is manufactured from china clay instead of ordinary clay which contains impurities such as iron oxide.

Coloured Portland cements are made by grinding the white cement with suitable pigments to produce the pastel shades. Darker shades allow the use of O.P.C. with the pigments.

These cements are used for decorative purposes and for roads and floors.

Hydrophobic cement

Hydrophobic cement is prepared by grinding Portland cement with substances such as oleic or stearic acids which are related to the soaps. Each particle is thus enclosed within a water-repellent film which protects it during storage under adverse conditions but is easily broken up during mixing.

Setting and hardening of Portland cement

A cement paste first stiffens to form a rigid mass with little strength and then slowly acquires mechanical strength. The first stage is known as setting and the second is known as hardening.

Setting is, in practice, divided into two parts, the initial and final setting times being times to reach arbitrarily chosen degrees of stiffness as defined in B.S.12. This is dealt with later in this chapter in the section on cement testing. Portland cement is not one chemical compound but rather a mixture of several complex entities. The four major constituents are listed in Table 7. During setting these combine with water by chemical processes which are covered by the general term 'hydration'. Hydration is exothermic, accounting for the observed temperature rise during setting. The setting is retarded by low temperatures and so in cold weather special precautions must be observed to conserve the heat.

The precise nature of the hardening process is, as yet, not completely understood. It appears that the products of hydration form into minute particles called 'colloids' which have, in consequence of their small size, a large surface/volume ratio. The mass of colloidal hydrates is called a 'gel' and includes some crystalline calcium hydroxide formed during the hydration of the silicates and aluminates. The gel contains voids which, again, are very small and are called 'interstitial' voids. These interstitial voids allow water to move between the capillary pores in the cement paste and the surfaces of the colloidal particles. The water/cement ratio is one of the factors determining the porosity of concrete; this point

c

will be considered in greater detail in Chapter Five. The strength
arises from the interlocking of the hydrate crystals and also from
the attraction between the closely packed particles.

Raw materials

Name of compound	Approximate percentage	Effect on finished cement
Lime	65	A high proportion gives a high early strength and increases the setting time. Insufficient causes a reduction in strength. Unsoundness arises from hard burnt excess lime
Silica	20	A high proportion gives both a high strength and longer setting time
Alumina	5	A high proportion gives a lower strength and shorter setting time
Iron oxide	3	Responsible for the grey appearance

Finished cement powder

Name of compound	Abbreviation	Approximate percentage	Effect on finished cement
Tricalcium silicate	C_3S	55	Produces much heat and early strength. A greater proportion in rapid-hardening cement than in low-heat cement
Dicalcium silicate	C_2S	17	Slower hydration and less heat with later strength development
Tricalcium	C_3A	11	Rapid reaction with water with high heat production. Responsible for *flash setting* if gypsum omitted. Attacked by sulphates and so less is contained by sulphate-resisting cements
Tetra-calcium alumina ferrite	C_4AF	9	Acts as a flux in manufacture. Produces little heat or strength. Acts as a flux assisting the manufacturing stage

Table 7   Main constituents of cement

Other cements

Aluminous cement (e.g. 'Ciment Fondu') (B.S. 915:1947)
Aluminous cement is manufactured by heating limestone and bauxite (an aluminium ore consisting mainly of hydrated alumina, oxides of iron, and silica) in an electric furnace. Water and carbon dioxide are driven off and at 1600°C the materials fuse together. The melt is run off and formed into *pigs* which are then broken down and ground to a fine powder.

The main hardening process is the hydration of calcium aluminate ($CaO.Al_2O_3$) to form calcium aluminate hydrate. The final strength is virtually gained within a few days whereas Portland cement gains in strength quickly at first but the process continues over a period of years. However, if exposed to warm moist conditions aluminous cement concrete is considerably weakened (*reversion*). Owing to the heat generated this cement must be kept wet whilst curing and placed in thin sections. The initial set occurs between 2 and 6 hours after mixing and the final set is within 2 hours of the initial set.

It has good sulphate resistance but is attacked by caustic alkalis. When used with crushed firebrick aggregate it is stable at temperatures up to 1300°C and with special aggregates 1800°C can be tolerated; so it is widely used as a refractory concrete. Portland cement, in contrast, breaks down if exposed to temperatures greater than 500°C for an extended period.

Supersulphated cement
When blast-furnace slag (80 to 85%), burnt gypsum or anhydrite (10 to 15%), and Portland cement (5%) are mixed and ground a cement is produced which is extremely resistant to sea-water, sulphates, acids found in peaty soils, and oil. It has a low heat of hydration (the Portland cement is added as an activator), should not be mixed with other cements, and requires wet curing.

It is used in sewer construction.

Tests for Portland cement (ordinary and rapid hardening)

The testing of Portland cement (ordinary and rapid hardening) is covered by B.S. 12:1958 to which reference must be made if greater detail is required.

The tests to be carried out are:
fineness
chemical composition
strength
setting time
soundness

The cement should comply with the limits set out in Table 8.

Fineness	Ordinary 225 m²/kg	Rapid hardening 325 m²/kg
Compressive strength (min.)		
(a) 3 days—		
Mortar cubes	15·16 N/mm²	20·68 N/mm²
Concrete cubes	8·27 N/mm²	11·72 N/mm²
(b) 7 days*—		
Mortar cubes	23·44 N/mm²	27·58 N/mm²
Concrete cubes	13·79 N/mm²	17·24 N/mm²
Tensile strength		
(optional) 24 hours		2·07 N/mm²
Setting time		
(a) Initial (min.)	45 min	45 min
(b) Final (max.)	10 h	10 h
Soundness (expansion)		
(a) Initial test (max.)	10 mm	10 mm
(b) Repeat test (max.)†	5 mm	5 mm

* The 7 day test result must exceed that of the 3 day test.
† The repeat test is carried out on a sample aged in air (R.H. 50 to 80%) for 7 days in a layer 75 mm deep. It should fail a cement which was unsound according to the initial test.

Table 8   Requirements for ordinary and rapid-hardening
Portland cement (excluding chemical composition)

Fineness
It is necessary to specify limits for the fineness of cement since it affects the rate of hydration. Hydration commences at the surface of the particles and, since finer cement presents a greater surface area to the water, it will harden more quickly. The fineness also affects the rate of deterioration on exposure to the atmosphere, the reaction with certain aggregates, and the texture of the finish.

Fineness is measured by *specific surface* which is the total surface area of all the particles in a given weight of cement expressed in m²/kg. Attempts to measure fineness by sieving have proved abortive owing to clogging of the extremely fine meshes needed and so B.S. 12 prescribes the use of the Lea and Nurse permeability apparatus (Fig. 9).

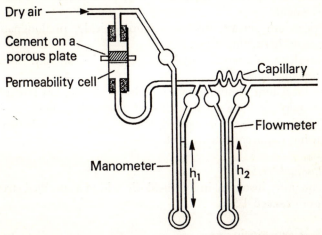

Fig 9   Permeability apparatus for the determination of fineness
(specific surface)

A weight of cement is chosen to give a porosity of 0·475 in a bed 10 mm thick in the permeability cell. This weight is found from a table in B.S.12 and depends on the density of the cement. A constant flow of dry air, measured by the flowmeter, is passed through the apparatus and the pressure drop across the bed is indicated by the manometer.

For a given porosity (0·475 in this case) the specific surface, S_w, is given by—

$$S_w = \frac{K}{p}\sqrt{h_1/h_2}$$

where K = a constant for the apparatus used.
   h_1 = difference in manometer levels.
   h_2 = difference in flowmeter levels.

Thus the fineness is measured by determining the resistance offered by a given volume of the cement to the passage of a stream of air under standard conditions.

Chemical composition
Portland cement does not have a fixed composition but is a mixture
of complex compounds the relative amounts of which depend on
the composition of the raw materials and the conditions of manu-
facture. Four compounds are accepted as being the major con-
stituents:
   Dicalcium silicate—hydrates slowly, predominates in low heat
      cement.
   Tricalcium silicate—hydrates more quickly, predominates in
      rapid hardening cement.
   Tricalcium aluminate—hydrates quickly, low strength, attacked
      by sulphates.
   Tetracalcium aluminoferrite—hydrates slowly, low strength.
There may also be a small residue of uncombined lime.
   Chemical analysis is carried out to determine whether the
following requirements are met.

(a) *Insoluble residue*
The quantity insoluble in hydrochloric acid of a specified strength
shall not exceed 1·5%.

(b) *Loss on ignition*
The total loss on ignition shall not exceed 3%.

(c) *Magnesia*
The magnesia content shall not exceed 4%.

(d) *Alumina/iron ratio*
The ratio of aluminium oxide to iron oxide shall not exceed
0·66%.

(e) *Lime saturation factor*
The lime saturation factor
$$=\frac{CaO - 0·7\ SO_3}{2·8\ SiO_2 + 1·2\ Al_2O_3 + 0·65\ Fe_2O_3}$$
shall not exceed 1·02 or be less than 0·66.

(f) *The total sulphur content,* expressed as sulphuric anhydride
(SO_3), shall not exceed 2·5% when the tricalcium aluminate content does not exceed 7%; otherwise it shall not exceed 3%.

Compressive strength
There are two standard methods detailed in B.S.12 for testing the compressive strength of cement. One uses mortar and the other concrete.

EXPERIMENT 15
To measure the compressive strength of cement using the mortar test.

Apparatus
Moulds (6) (cubical, 5000 mm² face)
Base plates
Grease
Trowel
Vibrating table
Cement
Standard sand
Compressive strength machine

Procedure
(a) Using the proportions:
Cement   185 g
Sand   555 g
Water   74 g
mix the cement and sand dry with a trowel for 1 minute and then with the water for 4 minutes.
(b)   Grease the moulds and base plates.
(c)   Fill the moulds.
(d)   Compact on the vibrating table for 2 minutes as 12000 vibrations per minute.
(e)   Cure at 20°C in a relative humidity of not less than 90% for 24 hours—covered.
(f)   Remove from the moulds and submerge in water at 20°C.
(g)   Determine the compressive strength of 3 cubes after 3 days.
(h)   Determine the compressive strength of a further 3 cubes after 7 days.

Conclusion
(a)   State whether the cement complies with B.S.12 taking the average value of the 3 cubes in each case.
(b)   Obtain the results from other groups in the class and calculate the standard deviation.

Concrete test

The concrete test requires rather more experience on the part of the operator if reliable results are to be obtained. Cement and water are mixed with a predetermined amount of coarse and fine aggregate to a standard degree of *workability* as measured by the slump test which is described below. The water/cement ratio stipulated is 0·6 and the slump between 12 and 50 mm. The mix is then cast into 101·6 mm cubes, aged, and tested.

Slump test

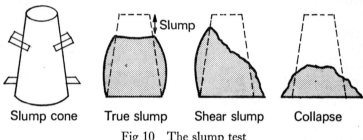

| Slump cone | True slump | Shear slump | Collapse |

Fig 10   The slump test

The term workability is used to describe the ease with which the concrete can be compacted. The slump test has been in use on sites for many years but for greater accuracy the compacting factor test (p. 118) is used. The slump test is nevertheless extremely useful in detecting variations in the uniformity of mixes. Both tests are described in B.S. 1881 'Methods of testing concrete'.

The mould for the slump test is shown in Fig 10. It is placed on a smooth surface with the larger end at the bottom and filled with concrete in four layers. Each layer is tamped 25 times with the tamping rod which is a 610 mm long steel rod, diameter 16 mm, with a rounded end. The cone is then lifted vertically leaving the concrete unsupported on the smooth surface. The decrease in height of the concrete as it slumps is measured. The test should be repeated if the slump is not *true*, that is the concrete has not slumped evenly but has either slid down an inclined plane (shear slump) or even collapsed completely (Fig 10).

EXPERIMENT 16

To measure the compressive strength of cement using the concrete test.

Apparatus

Moulds (6) (cubical, 101·6 mm side)

Slump cone

Tamping bar (steel bar weighing 1·814 kg, 381 mm long, ramming face 25·4 mm square)

Non-porous plate

Grease

Trowels (2)

Compression testing machine

Cement

Aggregate:

(a) Coarse. This should conform to B.S. 882. It should all pass a 19 mm (B.S. $\frac{3}{4}$ in) sieve and be substantially retained on a 4·8 mm (B.S. $\frac{3}{16}$ in) sieve. The amount passing a 200 mesh sieve should not exceed $\frac{1}{2}\%$.

(b) Fine. Siliceous sand conforming to B.S. 882. It should all pass a 4·8 mm (B.S. $\frac{3}{16}$ in) sieve. Not more than 10% should pass a 100 mesh sieve and not more than 2% should pass a 200 mesh sieve.

Both types of aggregate should be dry.

Procedure

(a) For each cube weigh out:

Cement: 325 g.

Water: 195 g.

Aggregate: determine the proportions by trial and error to achieve a slump between 12 and 50 mm.

The water/cement ratio is 0·6. As a guide to the proportions of aggregate to use start with a cement:fine aggregate:coarse aggregate = 1:2:4.

(b) Mix the cement and sand dry for 1 minute using two trowels.

(c) Add the coarse aggregate and mix to give a uniform colour.

(d) Mix in the water with two trowels for 3 minutes.

(e) Grease the moulds.

(f) Fill the moulds in two layers tamping each layer 35 times.

(g) Remove air bubbles by passing the blade of the trowel between the concrete and the mould wall.

(h) Smooth off the top with a trowel.

(i) Cover and leave in an atmosphere of not less than 90% R.H. and at a temperature of 19°C for 24 hours.

(j)   Mark each cube to identify it, and submerge in water at 19°C after removing it from the mould.
(k)   After 3 days remove three cubes for testing in the compression testing machine.
(l)   After 7 days test the remaining three cubes.

Conclusion
(a)   State whether the cement complies with B.S. 12 taking the average value of the three cubes in each case.
(b)   Obtain the results from other groups in the class and calculate the standard deviation.

Setting time
It was mentioned earlier in this chapter that the initial and final setting times are the times for the cement to reach arbitrarily chosen degrees of stiffness as defined in B.S. 12. A paste of standard consistence is prepared using the Vicat apparatus shown in Fig 11. The same apparatus is used for determining the setting times. The procedure is given in Experiment 17.

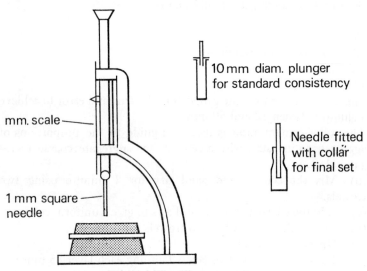

Fig 11   The Vicat apparatus

EXPERIMENT 17
To prepare a cement paste of standard consistence and measure the initial and final setting times.

Apparatus
Vicat apparatus, mould, plunger, needle, and annular attachment.
Cement.

Procedure

A   *Preparation of a cement paste of standard consistence*
(a)   Mix a trial paste of cement and water for not more than 4
minutes.
(b)   Fill the Vicat mould with the paste and smooth off.
(c)   Fit the Vicat apparatus with the 10 mm diameter plunger.
(d)   Place the filled mould under the plunger and lower the latter
into contact with the surface of the paste. Release the plunger.
Note how far the plunger is from the bottom of the mould when
it finally comes to rest.
(e)   Repeat with other trial mixes varying the water/cement ratio,
until the plunger is permitted to settle 5 to 7 mm from the bottom
of the mould. This is the paste of standard consistency.

B   *Determination of initial setting time*
(a)   Fit the 1 mm square needle to the Vicat apparatus.
(b)   Fill the mould with the paste of standard consistency and
place under the needle.
(c)   Lower the needle into contact and release.
(d)   Note the time elapsing between when the water was added
and when the needle fails to penetrate beyond a point approximately 5 mm from the bottom of the mould. This is the initial
setting time.

C   *Determination of final setting time*
(a)   Fit the needle with the annular attachment.
(b)   Note the time from adding the water to when only the needle
makes an impression, and the attachment fails to do so. This is the
final setting time.

Conclusion
State whether the cement conforms to B.S. 12.

Soundness
The soundness test examines the expansion occurring in the set
cement paste.

The particular type of unsoundness tested for by the method specified in B.S. 12 is that due to particles of free lime remaining in the clinker when the cement is manufactured. If, during the manufacture, more lime has been used than can combine with the other minerals then the excess appears in the cement clinker as hard burnt quicklime. This hydrates very slowly, the product having a greater volume than the quicklime. This is similar to the unsoundness occurring in lime discussed in Chapter One.

Under the conditions of use the expansion of concrete due to unsoundness is very gradual. It may take years to become noticeable by which time serious damage may have been caused to the structure. Under test conditions the test piece is aged in boiling water, so accelerating the expansion that it can be detected in 24 hours. A cement paste of standard consistence is placed in the cylinder of the Le Chatelier apparatus (p. 18) and aged under standard conditions, and the increase in the separation of the points is noted; this should not exceed 10 mm. If the cement fails another sample should be taken from elsewhere in the batch, aerated for 7 days, and tested. The expansion in this case should not exceed 5 mm.

EXPERIMENT 18
To test a sample of Portland cement for soundness.

Apparatus
Le Chatelier mould
Glass plates (2)
Weight (100 g)
Grease
Metal rod (for tamping)
Boiler
Cement
Vicat apparatus

Procedure
(a)   Prepare a paste of standard consistence using the Vicat apparatus as in Experiment 17.
(b)   Grease the Le Chatelier mould and place it on a greased plate.
(c)   Fill the mould, tamping to avoid the inclusion of air, and holding the split closed.

(d)   Cover the mould with a glass plate. Hold the plate down by means of the 100 g weight.
(e)   Submerge in water at 19°C for 24 hours.
(f)   Measure the separation of the points.
(g)   Submerge in water and bring to the boil in 25 to 30 minutes.
(h)   Boil for 1 hour.
(i)   Allow to cool.
(j)   Measure the separation of the points.

Conclusion
State whether the cement complies with B.S. 12.

Questions
1   Outline the manufacturing process for ordinary Portland cement. Why is gypsum added and at what stage?
2   Name *three* types of Portland cement and account for the differences in their properties. For what purpose is each particularly suitable?
3   Differentiate between 'setting' and 'hardening' of cement giving a brief description of the mechanism by which each is thought to occur.
4   How is aluminous cement manufactured? Compare the properties of aluminous cement with those of ordinary Portland cement. For which applications would you prefer aluminous cement?
5   List the British Standard tests to be applied to ordinary Portland cement. Where would you find full details of them? Why is each one of them important?
6   How would you carry out the following tests on cement (O.P.C.)?
(a)   Setting time.
(b)   Soundness.
Sketch the apparatus used and give the test limits.

Four Aggregates

Aggregates are materials such as sand and gravel which are dispersed throughout a cement paste and, when the cement sets and hardens, become part of a cohesive mass. In this chapter particular attention will be given to aggregates for concrete but they are also used in conjunction with other binders, notably gypsum plaster.

In some cases the chemical properties of the aggregate influence the properties of the final product but, more usually, an intert material is chosen in which case the physical properties are all-important. An aggregate may be added to cement for any of the following reasons:

(a) Economy—Aggregate is cheaper than cement and may be regarded as an extender.

(b) Durability—Incorporation of a suitable aggregate increases the resistance of the product to weathering and abrasion.

(c) Volume stability—Aggregates restrain the amount of shrinkage; it is found that a large aggregate permits the use of a leaner mix of concrete and so reduces the drying shrinkage. For a given mix strength (water/cement ratio) an increase in the aggregate proportion reduces the shrinkage. The expansion on setting is also reduced by the presence of aggregate.

(d) Adjustment of density—Aggregates are available in a wide range of densities making possible the production of concrete of different densities.

(e) Special characteristics—Special aggregates are available which modify the thermal insulation, acoustic, and fire resistance characteristics of the concrete. The appearance, colour, and texture of the concrete may also be modified.

The term 'aggregate' is reserved for material having a particle size greater than about 0·07 mm and tests must be carried out to ensure that it does not contain excessive quantities of material finer than this. Material having a particle size between 0·07 and 0·002 mm is called *silt*, and *clay* is smaller still. Loam contains clay, silt, and the smallest sized aggregate; it is rather soft. These fine particles, and fine dust created during the crushing of stone, may be present in sufficient quantity to coat the surface of the aggregate and interfere with the bond which develops between the aggregate and cement paste. The strength and durability of the concrete will be impaired. The strength of the product is also dependent on the strength and shape of the aggregate; the latter feature is important since it affects the degree of interlocking between the particles.

Classification of aggregates

Aggregates are divided into three classes depending on their relative density (specific gravity).

(a) *Normal aggregates*

Normal weight aggregates have relative densities between 2·5 and 3·0 and are generally obtained from natural sources such as sand, gravel, or crushed stone. They can also be manufactured from industrial products notably blast-furnace slag. Crushed brick is sometimes used and confers good fire resistance although, owing to its porous nature, it should not be used where low permeability is required. It also produces concrete having low abrasion resistance.

(b) *Lightweight aggregates*

The low density of lightweight aggregates stems from their high porosity. Concrete containing these aggregates has densities in the range 320 to 1900 kg/m^3 compared with the 2200 to 2600 kg/m^3 of other concretes. This makes it possible to use smaller supporting sections and foundations but it must be realised that the presence of the voids reduces the strength of the concrete and so it should only be used where strength is not important. It has sufficient durability for most purposes but its abrasion resistance is low; it is more expensive than normal concrete and requires more care in mixing, handling, and placing. It is usually rendered when used in exterior surfaces. Lightweight concrete and plasters are particularly valuable where improved thermal insulation is required. The thermal conductivity of lightweight concrete can be as low as

0·09 W/m°C compared with 1·4 to 3·6 W/m°C for other concretes.

Some lightweight aggregates are manufactured whilst others occur naturally. Those in common use are listed below.

Clinker

This is the residue from furnaces burning coke or coal. It should be well burnt since if combustible material remains it is liable to cause cracking eventually. Breeze and clinker have a high sulphur content and should not be used in reinforced concrete.

Foamed slag

This is manufactured by rapidly quenching blast-furnace slag with water in such a manner as to encourage the formation of pores. It is used in both partitions and external concrete walls.

Exfoliated vermiculite

Vermiculite particles, a form of mica, are expanded by heating to 600 to 900°C. The product is used in plaster, partitions, and insulation slabs for floors.

Expanded perlite

Perlite, a glassy volcanic rock, produces a cellular aggregate when heated. It is used in lightweight plasters.

Expanded shale and clay

Both these materials can be expanded by heating to a high temperature. Expanded shale imparts higher strength, lower drying shrinkage, and reduced moisture movement than other lightweight aggregates.

Pulverised fuel ash

Fly ash from power stations is made into porous pellets by mixing with water and sintering in an autoclave at 1200°C. Its performance in concrete is similar to that of clinker.

Pumice

Pumice is a glassy rock of volcanic origin and cellular structure. It is used for precast concrete partition slabs, encasement of steelwork, and insulating screeds.

(c)   *Heavy aggregates*

The heavy aggregates commonly used are barytes, iron ore such as magnetite, and steel punchings. The high density concrete produced from them is used in the construction of screens for radioactive sources and atomic reactors.

British Standards

Frequent reference will be made in this chapter to two important British Standards covering aggregates. They are B.S. 812:1967 and B.S. 882:1965.

B.S. 812. 'Methods for sampling and testing of mineral aggregates sands and fillers'
This British Standard specifies methods for the sampling, classification, and testing of mineral aggregates, sands, and fillers. It also includes a classification of stone according to petrological characteristics (mode of origin, chemical and mineral composition, and present condition, e.g. fused, granular, or crystalline) and a glossary of rock names.

The tests cover:
Particle size
Particle shape
Specific gravity and water absorption
Density
Voids
Bulking
Moisture content
Detection of organic impurities
Determination of mechanical properties: resistance to impact, crushing, abrasion, and polishing.

B.S. 882. 'Specification for aggregates from natural sources for concrete'
This British Standard relates to naturally occurring materials, crushed or uncrushed, used in the production of concrete for normal structural purposes, including roads.

Associated with B.S. 882 is B.S. 1201:1965 'Aggregates for granolithic concrete floor finishes'.

Quality of aggregates

It is specified in B.S. 882 that the aggregates shall be hard, durable, and clean, and shall not contain deleterious materials in such a form or in sufficient quantity to affect adversely the strength at any age or the durability of the concrete, including, where applicable, the resistance to frost and to corrosion of the reinforcement. Examples of such materials are:

Clay, particularly as an adherent coating

Flaky or elongated particles

Mica, shale, and other laminated materials

Coal and other organic impurities

Iron pyrites and soluble sulphate salts such as those of calcium, magnesium, and sodium.

Earlier in this chapter it was mentioned that impurities such as clay affect the bond development, owing to providing a loosely adherent layer on the aggregate particles. Clay and salts may also affect the rate of hardening of the concrete and even attack the cement. The salts may be drawn in solution to the surface of the concrete and be deposited as an unsightly deposit known as 'efflorescence'. Coal may swell and disrupt the concrete besides affecting the hardening process. Iron pyrites react with oxygen and water in the atmosphere to form undesirable sulphates capable of attacking the cement. Surface staining is also likely.

Sampling of aggregates

In order that the aggregate may be tested it is necessary to obtain a sample of suitable size which is truly representative of the stock-pile.

The main sample consists of ten portions drawn by means of a scoop from different parts of the bulk quantity. The sample should be, as far as possible, representative of the average quality.

Smaller portions of this main sample are taken for testing using a sample divider (Fig 12) or by quartering.

(a) *Use of a sample divider*

When the sample is passed through the riffle box half falls to one side and half to the other. One half is passed through again and the process is repeated until a sample of appropriate size is obtained.

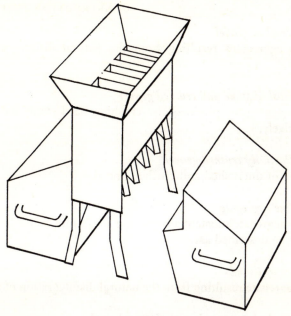

Fig 12   Riffle box

(b) *Quartering*
The main sample is heaped to form a cone which is then turned
over using a shovel to form a new cone. This is repeated twice
more before the cone is flattened, divided into four equal parts,
and two diametrically opposite portions are removed. The process
is repeated until the required weight of sample is obtained.

Particle size distribution—grading of aggregates

The determination of the particle size distribution of aggregates is
called 'grading'. It is carried out by shaking the sample on a series
of sieves of standard mesh sizes (B.S. 410:1961 'Specification for
Test Sieves'). The result is expressed as the percentage of sample
weight retained on each sieve commencing with the largest size.
The maximum and minimum weights of sample to be taken are
specified in B.S. 812 which also defines aggregate as being *coarse,
fine,* or *all-in.*

(a) *Coarse aggregate*
This is aggregate substantially retained on a 4·76 mm coarse sieve.
It may be described as:

(i) *Uncrushed gravel*
Coarse aggregates resulting from the natural disintegration of rock.

(ii) *Crushed stone and crushed gravel*
Coarse aggregates produced by crushing hard stone and gravel respectively.

(iii) *Partially crushed gravel*
A blend of uncrushed and crushed gravel.

(b) *Fine aggregate*
This is aggregate substantially passing a 4·76 mm coarse test sieve. It may be described as:

(i) *Natural sand*
Fine aggregate resulting from the natural disintegration of rock.

(ii) *Crushed stone and crushed gravel sand*
Fine aggregate produced by crushing hard stone and gravel respectively.

(c) *All-in aggregate*
This is material composed of a mixture of coarse aggregate and fine aggregate.

EXPERIMENT 19
To determine the particle size distribution of a coarse aggregate by sieve analysis.

Apparatus
B.S. sieves (300 mm diameter)—
   38·1 mm mesh
   19·05 mm mesh
   9·52 mm mesh
   4·76 mm mesh
Sieve lid
Sieve pan
Vibrating machine
Dry coarse aggregate (19 to 5 mm)

Procedure

(a)   Arrange the sieves on the vibrating machine in order of size placing the sieve pan on the bottom and the largest mesh sieve on the top.

(b)   Weigh out between 2 kg and 3 kg of aggregate to the nearest 2 g and place it on the top of the nest of sieves.

(c)   Fit the lid and vibrate for 20 minutes.

(d)   Weigh the material retained on each sieve and that which is collected in the pan.

Results

(a)   Tabulate the results as follows.

B.S. test sieve	Material retained		Material passing (to nearest whole number)
	g	%	
38·1 mm			
19·05 mm			
9·52 mm			
4·76 mm			
Passing 4·76 mm			
Total			

(b)   Prepare a *grading chart*. Connect the points by straight lines.

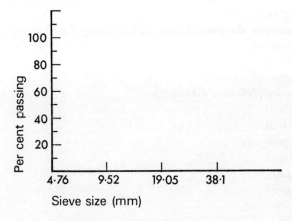

(c)   Calculate the *fineness modulus* (F.M.) (p. 72).

$$\text{F.M.} = \frac{\text{Sum of cumulative \% retained on each sieve}}{100}$$

Conclusion

State whether the material is within the limits specified in Table 9 for:

(a)   Graded aggregate.
(b)   Single-sized aggregate.

| Test mm | Percentage by weight passing B.S. sieve | | | | | | | |
| | Nominal size of graded aggregate | | | Nominal size of single sized aggregate | | | | |
	38 mm to 5 mm	19 mm to 5 mm	13 mm to 5 mm	64 mm	38 mm	19 mm	13 mm	10mm
76·20	100	—	—	100	—	—	—	—
63·50	—	—	—	85–100	100	—	—	—
38·10	95–100	100	—	0–30	85–100	100	—	—
19·05	30–70	95–100	100	0–5	0–20	85–100	100	—
12·70	—	—	90–100	—	—	—	85–60	100
9·52	10–35	25–55	40–85	—	0–5	0–20	0–45	85–100
4·76	0–5	0–10	0–10	—	—	0–5	0–10	0–20
2·40	—	—	—	—	—	—	—	0–5

Table 9   Limits of grading for coarse aggregate (B.S. 882)

EXPERIMENT 20
To determine the particle size distribution of a fine aggregate by sieve analysis.

Apparatus

B.S. sieves (200 mm diameter)—
    9·52 mm
    4·76 mm
    2·40 mm
    1·20 mm
    600 μm
    300 μm
    150 μm
Sieve lid
Sieve pan
Vibrating machine
Dry sand

Test sieve mm	Percentage by weight passing B.S. sieve			
	Grading Zone 1	Grading Zone 2	Grading Zone 3	Grading Zone 4
9·52	100	100	100	100
4·76	90–100	90–100	90–100	95–100
2·40	60–95	75–100	85–100	95–100
1·20	30–70	55–90	75–100	90–100
0·600	15–34	35–59	60–79	80–100
0·300	5–20	8–30	12–40	15–50
0·150	0–10	0–10	0–10	0–15

Table 10   Grading zones for fine aggregate (B.S. 882)

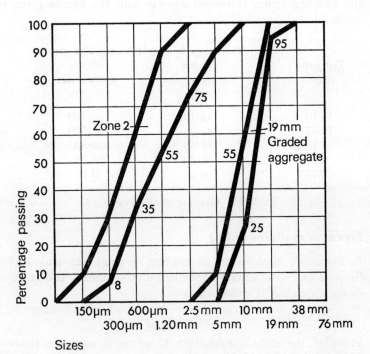

Fig 13   B.S. grading limits for Zone 2 fine aggregate and for
19 mm graded aggregate

Procedure

Follow the procedure given in Experiment 19 using a 200 g sample.

Results

(a)   Tabulate the results as for Experiment 19.

(b)   Prepare a grading chart.

(c)   Insert on the grading chart the zone boundaries given in Table 10.

(d)   Calculate the fineness modulus as described in Experiment 19.

Conclusion

State the grading zone into which the aggregate falls, using Table 10 (see Fig 13).

All-in aggregate is graded by first separating it into two fractions using either a 3·18 mm or a 4·76 mm sieve whichever is the more convenient. The two fractions are then graded separately as coarse and fine aggregate. It should comply with the figures given in Table 11.

| | *Percentage by weight passing B.S. sieves* | |
Test sieve *mm*	*38 mm* *nominal size*	*19 mm* *nominal size*
76·20	100	—
38·10	95–100	100
19·05	45–75	95–100
4·76	35–45	30–50
0·600	8–30	10–35
0·150	0–6	0–6

Table 11   Grading of all-in aggregate

Fineness modulus (F.M.)

In the above experiments on grading mention was made of the fineness modulus. This is dependent on the number and range of sieves used since it is defined as—

$$\text{F.M.} = \frac{\text{Sum of cumulative \% retained on each sieve}}{100}$$

Provided the same combination of sieves is used the fineness modulus is valuable for routine checking on variations in the aggregate supply and for proportioning aggregates. (*See* also p. 94.)

Example 1   An aggregate has the following grading:

B.S. test sieve mm	Percentage passing
9·52	100
4·76	98
2·40	95
1·20	90
600 μm	67
300 μm	40
150 μm	8

Calculate the fineness modulus for the range of sieves used.
The first step is to calculate the cumulative percentage retained.

B.S. test sieve mm	% passing	Cumulative % retained
9·52	100	0
4·76	98	2
2·40	95	5
1·20	90	10
600 μm	67	33
300 μm	40	60
150 μm	8	92
		Total 202

Then—

$$\text{F.M.} = \frac{\text{Sum of cumulative \% retained}}{100}$$

$$= \frac{202}{100} = 2{\cdot}02$$

Ans. The fineness modulus is 2·02.

Example 2   An aggregate is required which has a fineness modulus of 4·5. What weight of coarse aggregate having a fineness modulus of 6·1 must be mixed with each 100 kg of fine aggregate if the latter has a fineness modulus of 2·1?
Let W kg = the weight of coarse aggregate required

Then—
$$(100 \times 2 \cdot 1) + (W \times 6 \cdot 1) = (100 + W) \times 4 \cdot 5$$
$$210 + 6 \cdot 1 \ W = 450 + 4 \cdot 5 W$$
$$W = 150$$

Ans. The weight of coarse aggregate required is 150 kg.

Specific surface

Specific surface has already been mentioned in connection with cement when it was defined as the surface area of all the particles in a given weight. The term is also applied to aggregates. It is inversely proportional to the particle size—the finer aggregates have the greater specific surface—and influences the proportion of water that must be incorporated in a concrete mix. This will be mentioned again in the next chapter.

Gap-graded aggregate

Gap-graded aggregate has certain intermediate sizes missing. This is in contradistinction to continuously graded aggregate. The grading chart shows a horizontal line over the missing range.

The workability of concrete is discussed in the next chapter but it is appropriate at this stage to mention that it is increased by the use of gap-graded aggregate. Otherwise it could be increased by using a greater proportion of coarse aggregate, a procedure which increases the liability to *segregation*, which refers to the shaking out of the smaller particles from the voids during transport or compaction. The use of gap-graded aggregate does not reduce this; it may even increase it in highly workable mixes, and so it is recommended that it be used only where the workability is low and compaction is by mechanical vibration.

The use of gap-graded aggregate is economical since a smaller number of stock bins is required.

Determination of clay, silt, and dust in fine or coarse aggregates

The importance of checking the clay, silt, and dust content of aggregates has already been mentioned. It was pointed out that their presence in excessive amounts may result in reduced adhesion within the concrete. Looking forward to Chapter 5 the presence of fine particles affects the workability of the mix; owing to their higher surface area more water is required to wet all the particles and this leads to a higher water/content ratio requirement.

The methods given in B.S. 812 are:

Field settling test

Decantation method ⎫
Sedimentation test ⎭ Laboratory tests

(a) *Field settling test*

The field settling test is an approximate method intended for routine checking. It gives a guide to the clay and fine silt content of natural sand and crushed gravel but is not applicable to crushed stone sands or coarse aggregates. If the test shows that the clay and silt content is greater than 8% by volume the laboratory tests should be carried out.

EXPERIMENT 21

To carry out the field settling test on a sample of natural sand.

Apparatus

Measuring cylinder (250 ml)
Measuring cylinder (100 ml)
Beaker (250 ml)
Rule
Salt
Natural sand

Procedure

(a) Prepare a 1% solution of salt in water.
(b) Pour 50 ml of the salt solution into the 250 ml measuring cylinder.
(c) Add the sand gradually until the volume of the sand is 100 ml.
(d) Add more salt solution until the volume is 100 ml.
(e) Shake the mixture vigorously until adherent clayey particles have been dispersed.
(f) Place on a level bench and tap gently until the surface of the sand is level.
(g) Allow to stand for 3 hours and measure the height of the sand and the height of the silt layer above it.

Results
Express the height of the silt as a percentage of the height of the sand.

Conclusion
State whether the sample should be submitted for a laboratory test.

(b) *Decantation method*
This is a method for determining the amount of material finer than a 75 micron (75 μm) sieve.

 B.S. 812 specifies the following maximum limits:
 (i) Coarse aggregate, 1% by weight.
 (ii) Natural sand or crushed gravel sand, 3% by weight.
 (iii) Crushed stone sand, 15% by weight.
 (iv) All-in aggregate: The above requirements shall apply in proportion to the amounts of the respective materials present.

EXPERIMENT 22
To determine the amount of material finer than a 75 micron sieve using the decantation method.

Apparatus
B.S. sieves—
 75 micron
 1·20 mm
Bucket (10 l capacity)
Ventilated oven
Trays to fit inside the oven
Aggregate

Procedure
(a) Dry sufficient of the sample in the oven at 105°C to give the required dry sample weight of Table 12.
(b) Wet the sieves on both sides and place the coarser sieve on top of the finer one.
(c) Place the sample in the bucket and cover with water.
(d) Agitate the bucket vigorously in order to bring the fine material into suspension.
(e) Immediately pour the wash water through the sieves avoiding, as far as possible, pouring coarse material on to them.

Nominal maximum size, mm	Minimum weight of sample, kg
63·50–25·40	6
19·05–12·70	1
9·52–6·35	0·5
4·76 or smaller	0·2

Table 12   Weight of sample for decantation method

(f)   Repeat the washing and decantation until the wash water is clear.

(g)   Return all the material retained in the sieves to the washed sample.

(h)   Dry the washed sample in the oven, cool, and weigh.

Results
Calculate the percentage, P, of material finer than a 75 micron sieve from the formula—

$$P = \frac{A-B}{A} \times 100$$

where A = weight (g) of oven-dried sample.

B = weight (g) of portion retained on the 75 micron sieve.

Conclusion
State whether the sample complies with the limits specified in B.S. 812.

(c)   *Sedimentation method*
The sedimentation method is a gravimetric method for determining the clay, fine silt, and fine dust, which includes particles up to 20 microns.

The sample is washed with sodium oxalate solution in a suitable container and a portion of the washings is removed using a special pipette described in B.S. 812. The portion is then dried in an oven and the residue weighed. The residue weight is then expressed as a percentage of the sample weight.

Specific gravity and water absorption

Aggregate is a porous material. Some of the pores can be seen with the naked eye but some are only a few microns in diameter. Again, some of the pores open on to the surface while others are completely inside the solid. These factors affect such properties of the concrete as the strength of the bond, its resistance to abrasion, and its freeze-thaw stability.

The presence of the pores make it necessary to be very careful in definining specific gravity.

(a) *Absolute specific gravity*
This is the specific gravity of the solid alone. If the aggregate is ground down to a powder fine enough to ensure the complete absence of pores then the specific gravity at a particular temperature is given by—

$$\text{S.G.} = \frac{\text{Weight of powder}}{\text{Weight of an equal volume of water}}$$

(b) *Apparent specific gravity*
If the pores in the aggregate are full of water it is said to be 'saturated and surface-dry'. When the aggregate has geen dried in an oven it is said to be 'oven dried'.

The apparent specific gravity is then

$$\text{S.G.} = \frac{\text{Weight of the oven-dried aggregate}}{\substack{\text{Weight of water occupying an equal volume} \\ \text{including the pores}}}$$

Briefly the B.S. 812 test consists of immersing the aggregate for 24 hours in water and then weighing it in water. This is followed by drying it with a cloth and leaving it in the atmosphere until all visible films of water are removed but it still has a damp appearance. This saturated surface-dry aggregate is weighed. It is then dried at 104°C for 24 hours and re-weighed.
Then—

$$\text{Apparent specific gravity} = \frac{C}{C - A}$$

where A = the weight of the saturated aggregate in water.
C = the weight of the oven-dried aggregate in air.

(c) *Specific gravity on an oven-dried basis*
This is given by—

$$\frac{C}{B-A}$$

where B = the weight of the saturated surface-dry aggregate in air.
A and C are as above.

(d) *Specific gravify on a saturated and surface-dried basis*
This is given by—

$$\frac{B}{B-A}$$

where A, B, and C are as above.

The water absorption is calculated from the results obtained during the determination of specific gravity.

$$\text{Water absorption (per cent of dry weight)} = \frac{100\ (B-C)}{C}$$

In the above method the aggregate is weighed in water whilst contained in a wire basket. For aggregates smaller than 9·52 mm it is more convenient to use the pycnometer method which is also given in B.S. 882 and forms the basis of Experiment 23.

EXPERIMENT 23
To determine the specific gravity and water absorption of an aggregate smaller than 9·52 mm.

Apparatus
Ventilated oven
Tray to fit inside the oven (approx. area 0·033 m²)
Bucket (10 l capacity)
B.S. sieves—
    1·20 mm
    750 μm
Pycnometer (Fig 14)
Air-tight container
Hair dryer
Aggregate

Procedure
(a)   Weigh out the aggregate test samp'e—
1 kg for sizes 9·52 mm to 4·76 mm

0·5 kg for sizes less than 4·76 mm.

(b) Wash to remove material finer than 750 μm.

(c) Transfer the washed aggregate to the tray, and decant the water.

(d) Cover with distilled water at 20°C and remove air bubbles by gentle agitation.

(e) Allow to stand for 24 hours.

(f) Decant the water.

(g) Dry, stirring at intervals, using warm air from the hair dryer, until the aggregate is free running and there is no sign of surface water.

(h) Weigh, saturated and surface-dry ($=A$ kg).

(i) Place the aggregate in the pycnometer and fill with distilled water. Remove air bubbles and top up to remove froth.

(j) Dry the outside of the pycnometer and weigh ($=B$ kg).

(k) Transfer the contents to the tray.

(l) Fill the pycnometer with distilled water, dry the outside, and weigh ($=C$ kg).

(m) Decant the water from the sample and dry on the tray in the oven at 105°C for 24 hours.

(n) Weigh the oven-dried aggregate ($=D$ kg).

Fig 14   Pycnometer for the specific gravity determination (B.S. 812)

Results

Specific gravity on an oven-dried basis

$$= \frac{D}{A-(B-C)}$$

Specific gravity on a saturated and surface-dried basis

$$= \frac{A}{A-(B-C)}$$

Apparent specific gravity $= \dfrac{D}{D-(B-C)}$

Water absorption (% by weight) $= \dfrac{100\,(A-D)}{D}$

Two determinations are carried out and the mean calculated.

Flakiness index

The flakiness index of an aggregate is the percentage by weight of particles in it whose least dimension (thickness) is less than three-fifths of their mean dimension. The test is not applicable to material passing a 6·35 mm sieve.

A summary of the procedure is as follows:

(a) Separate the sample into fractions by sieving on normal test sieves the sizes of which are given in Table 13.

Size of aggregate	
Passing B.S. sieve	*Retained B.S. sieve*
63·50 mm	50·80 mm
50·80 mm	38·10 mm
38·10 mm	31·75 mm
31·75 mm	25·40 mm
25·40 mm	19·05 mm
19·05 mm	12·70 mm
12·70 mm	9·52 mm
9·52 mm	6·35 mm

Table 13   Size fractions for flakiness index

(b) Gauge in turn for thickness using either the thickness gauge described in B.S. 812 or in bulk using special sieves having

D

elongated slots. The gauge and slots have a width three-fifths the mean sieve size.

(c)   Weigh the total amount of each fraction passing the thickness gauge or special sieve.

The flakiness index is the total weight of the material passing the various thickness gauges or special sieves, expressed as a percentage of the total weight of the sample gauged.

Elongation index

The elongation index of an aggregate is the percentage by weight of particles whose greatest dimension (length) is greater than $1\frac{4}{5}$ times their mean dimension. The test is not applicable to material passing a 6·35 mm sieve.

The sample fractions obtained as in the flakiness test are gauged in turn for length using the length gauge described in B.S. 812.

The elongation index is the total weight of the material retained on the various length gauges, expressed as a percentage of the total weight of the sample gauged.

Angularity number

Angularity, or absence of rounding, of the particles of an aggregate is a property which is of importance because it affects the ease of handling of a mixture of aggregate and binder (e.g. the workability of concrete) or the stability of mixtures that rely on the interlocking of the particles. The B.S. 812 method is intended for determining this property of an aggregate for mix design and research purposes.

The angularity number is a measure of relative angularity based on the percentage of voids in the aggregate after compaction in the prescribed manner.

The classification of roundness adopted in B.S. 812 is given in Table 14.

A summary of the procedure for the determination of angularity number is given below. It is based on the fact that the most rounded aggregates have about 33% voids and is defined as the amount by which the percentage of voids exceeds this value. Its value ranges from 0 to 12.

(a)   Obtain a test sample by taking the aggregate retained between

Classification	Description	Examples
Rounded	Fully water-worn or completely shaped by attrition	River gravel Sea sand
Irregular	Naturally irregular, or partly shaped by attrition and having rounded edges	Pit gravel
Flaky	Material of which the thickness is small relative to the other two dimensions	Laminated rock
Angular	Possessing well-defined edges formed at the intersection of roughly planar faces	Crushed rock and slag
Elongated	Material, usually angular, in which the length is considerably larger than the other two dimensions	—
Flaky and elongated	Material having the length considerably larger than the width, and the width considerably larger than the thickness	—

Table 14   Particle shape (B.S. 812)

the appropriate pair of B.S. perforated-plate test sieves selected from:

$$19{\cdot}05 \text{ mm and } 12{\cdot}70 \text{ mm}$$
$$12{\cdot}70 \text{ mm and } 9{\cdot}52 \text{ mm}$$
$$9{\cdot}52 \text{ mm and } 6{\cdot}35 \text{ mm}$$
$$6{\cdot}35 \text{ mm and } 4{\cdot}76 \text{ mm}$$

(b)   Dry for at least 24 hours in a ventilated oven at 105°C.
(c)   Place the sample in the metal cylinder specified in the B.S., tamping down in the standard manner and strike off level with the top.
(d)   Weigh the contents.
    The angularity is given by—

$$\text{Angularity number} = 67 - \frac{100W}{CG_A}$$

where W = mean weight of aggregate in the cylinder.
    C = weight of water required to fill the cylinder.
    G_A = specific gravity on an oven-dried basis of the aggregate.

Bulk density

The bulk density is the weight of material in a given volume (kg/m³). It depends on the density of packing of the aggregate particles which, in turn, depends on the grading and particle shape. When it is being determined therefore due regard must be paid to the degree of compaction.

B.S. 812 specifies two methods for the determination of bulk density; they are summarised below. The container mentioned is a metal cylinder having dimensions appropriate to the size of aggregate and specified in B.S. 812.

(a) *Compacted bulk density (not applicable to moist fine aggregate)*
  (i) Fill the container $\frac{1}{3}$ full using a scoop discharging from a height not greater than 50 mm above the top of the container.
 (ii) Tamp the aggregate the specified number of times; this varies with the size of the aggregate.
(iii) Fill the container to overflowing, tamp again, and level off.
 (iv) Weigh the contents.

(b) *Uncompacted bulk density*
The method is the same as for the compacted bulk density except that the compaction with the tamping rod is omitted.

Calculation of voids and bulking

The percentage voids is calculated as follows:

$$\text{Percentage voids} = 100 \times \frac{(G_A \times 1000) - D_B}{(G_A \times 1000)}$$

where G_A = specific gravity, on an oven-dried basis.
          D_B = bulk density of the dried material (compacted or uncompacted, as required).

When water is added to an aggregate the volume increases owing to the film of water round the particles pushing them apart. This is called 'bulking' and depends on:
(a) The particle size. The degree of bulking is greatest for fine sand and negligible for coarse aggregate.
(b) The moisture content. The bulking increases up to about 30% as the moisture content of fine sand reaches about 8%. As the water content is increased further the bulking decreases. When the sand is fully saturated (inundated) the films of water round the particles have fully joined up, the voids are full, and the volume has dropped to that of dry sand.

The percentage bulking can be calculated from—

$$\text{Percentage bulking} = \frac{D_B(100 + n)}{D} - 100 \quad \text{(B.S. 812)}$$

where the symbols have the same meaning as given previously with the addition,

$D=$ uncompacted bulk density as determined at a moisture content of n per cent.

An alternative method of carrying out a bulking test is given in experiment 24.

EXPERIMENT 24
To carry out a bulking test on fine sand.

Apparatus
Gas jar
Tamping rod (400 mm × 10 mm diam.)
Metal tray
Rule
Fine sand (damp)

Procedure
(a)   Pour the damp sand into the gas jar until it is about ⅔ full.
(b)   Measure the depth of sand (H in Fig 15).

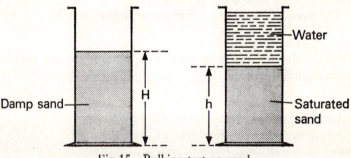

Fig 15   Bulking test on sand

(c)   Remove the sand onto the metal tray.
(d)   Half fill the glass jar with water and replace the sand carefully.
Use the tamping rod to ensure that no air bubbles are trapped.
(e)   Measure the depth of this saturated sand (h in Fig 15).

Conclusion
Calculate the percentage bulking from—

$$\text{Percentage bulking} = \frac{H-h}{h} \times 100$$

The influence of the water content of aggregate on the mix proportions of concrete

The moisture content of an aggregate has an influence on the mix proportions of concrete and similar compositions.

(a)   The weight of the water to be added must be reduced by the weight contained in the aggregate.

(b)   The weight of aggregate must be increased to allow for its water content.

(c)   When volume batching (proportioning by volume) is employed the apparent volume of aggregate must be increased to allow for bulking. This is particularly important when sand is being proportioned and neglect in this regard will result in a mix deficient in sand.

Determination of the moisture content of aggregate

The standard procedure for the determination of the moisture content of aggregate is the oven-drying method. This provides a measure of the total water present.

The oven-drying method is given as Experiment 25.

EXPERIMENT 25
To determine the moisture content of aggregate by the standard (oven-drying) method.

Apparatus
Balance (3 kg capacity, accurate to 0·5 g)
Air-tight non-corrodible container (3 kg capacity)
Scoop (200 mm long, 120 mm wide)
Well ventilated oven
Aggregate sample

Procedure
(a)   Weigh the clean dry container and lid ($=A$ kg).

(b)   Scoop about 2 kg of the aggregate into the container, replace the lid, and re-weigh ($=B$ kg).

(c)   Remove the lid and dry the container and its contents in the oven, controlled at 105°C, for 16 to 24 hours.

(d)   Remove from the oven, replace the lid, allow to cool for 1 hour, and re-weigh ($=C$ kg).

Results
Calculate the moisture content as follows—

$$\frac{B-C}{C-A} \times 100\% \text{ by dry weight}$$

or $\quad \dfrac{B-C}{B-A} \times 100\%$ by wet weight.

Three other methods are given in B.S. 812. They are:
 (i) A modified drying method. This method uses a shallow metal tray instead of the container mentioned in the experiment.
 (ii) Siphon-can method.
(iii) Buoyancy method.

The siphon-can method is suitable for field tests and is given as Experiment 26.

EXPERIMENT 26
To determine the moisture content of an aggregate by the siphon-can method.

Apparatus
Siphon-can (Fig 16)
Scales (to weigh 3 kg to the nearest 1 g; preferably with a scoop for one pan)
Weight (2 kg)
Measuring cylinder (500 ml)
Metal stirring rod (460 to 510 mm long, 6·5 mm diameter)

Procedure
A   *Calibration of the siphon can*
(a)   Wash out the can, drain the tubes, and tighten the screw clips.
(b)   Fill the can with clean water until the surface is above the upper tube.
(c)   Open the upper tube and discharge the water to waste.
(d)   Open the lower tube and run the water into the measuring cylinder.
(e)   Record the volume of water delivered ($= V$ ml).

B   *Routine procedure*
(a)   Add water to the can and discharge the excess to waste by opening the lower tube.
(b)   Close both screw clips.

(c)   Weigh out 2 kg from the sample of aggregate.

(d)   Transfer the weighed aggregate to the can taking care not to lose either aggregate or water in the process.

(e)   Wet the stirring rod and let excess water drain off.

(f)   Stir the aggregate to displace air bubbles.

(g)   Draw any scum on the surface of the water away from the centre to the side by means of the rod.

(h)   Remove the rod, letting excess water drain from the rod into the can.

(i)   Open the upper tube and discharge the water into the measuring cylinder.

(j)   Record the volume of water delivered ($=v_w$).

C   Preliminary test

Before the moisture content of the sample can be calculated it is necessary to make a preliminary test on a nominally similar sample of known moisture content because of the effect of the specific gravity of the aggregate parts on the value of v_w.

(a)   Obtain at least 5 kg of each aggregate in its normal damp condition and mix each sample thoroughly to ensure that its moisture content is uniform.

(b)   Weigh a 2 kg sample and make the siphon-can test in the usual way. Record the volume of water delivered as v_p.

(c)   Weigh out as soon as possible another 2 kg of sample and dry it—

If the total water content is required, dry completely.

If only the surface water is required, dry to a saturated surface-dry condition.

(d)   Almost fill the measuring cylinder with water and note the volume of water it contains.

(e)   Put the dried sample in the scoop of the balance and add water to it from the measuring cylinder until the weight of the aggregate and water is 2 kg.

(f)   Record the volume of water added to the aggregate ($=v$).

(g)   Calculate the value of v_b from the formula—

$$v_b = v_p - \frac{v}{2000 - v} \ (2000 - V - v_p)$$

Results

Moisture content may be expressed either by *dry weight* or by *wet weight* according to whether the weight of water in the sample of aggregate is divided by the weight of the aggregate dry or by the weight of the aggregate plus the water.

$$M_d = \frac{v_w - v_b}{2000 - V - v_w} \times 100 \text{ per cent}$$

$$M_w = \frac{v_w - v_b}{2000 - V - v_b} \times 100 \text{ per cent}$$

where M_d = moisture content by dry weight.

   M_w = moisture content by wet weight.

The other values are as given in the Procedure.

Alternatively the moisture content may be interpolated from tables given in B.S. 882.

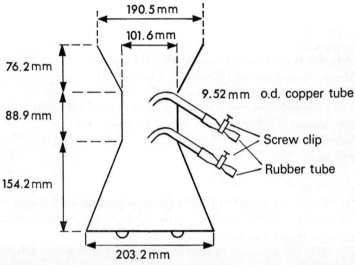

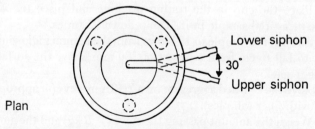

Fig 16   Siphon-can for determination of the moisture content (B.S. 882)

Determination of mechanical properties

Aggregate impact value (A.I.V.)
The aggregate impact value gives a relative measure of the resistance of an aggregate to sudden shock or impact. This value is not always the same as the resistance to a slow compressive load.

B.S. 882 specifies that, for concrete, the A.I.V. shall not exceed 45%. Where the concrete is for wearing surfaces the A.I.V. shall not exceed 30%.

EXPERIMENT 27
To determine the aggregate impact value.

Apparatus
Impact testing machine (Fig 17)
Cylindrical steel cup (51 mm deep, 102 mm diam.)
Metal tup or hammer (13·6 to 14·1 kg weight)
Cylindrical measure (51 mm deep, 76 mm diam.)
Tamping rod (230 mm × 10 mm diameter, rounded end)
B.S. sieves—12·70 mm, 9·52 mm, 2·40 mm
Balance
Aggregate

Procedure
(a)   Prepare a sample of the aggregate passing a 12·7 mm but retained on a 9·52 mm sieve, surface-dry. (Smaller sizes—use the appropriate sieves given in B.S. 812.)
(b)   Fill the measure to overflowing in 3 equal stages tamping each 25 times. Level off.
(c)   Weigh the contents of the measure ($= A$ kg).
(d)   Place the cup in the testing machine and place the whole of the measured sample in it. Tamp down 25 times.
(e)   Subject the sample to 15 blows from the hammer allowing the latter to fall freely from a height of 380 mm above the surface of the aggregate.
(f)   Sieve the crushed aggregate on a 2·40 mm sieve (or appropriate sieve with finer samples).
(g)   Weigh the amount passing the sieve ($= B$ kg) and the amount retained ($= C$ kg). (Repeat test if $B + C$ is less than A, the initial weight.)

Results

The aggregate impact value is given by—

$$A.I.V. = \frac{B}{A} \times 100$$

The mean should be calculated from two determinations.

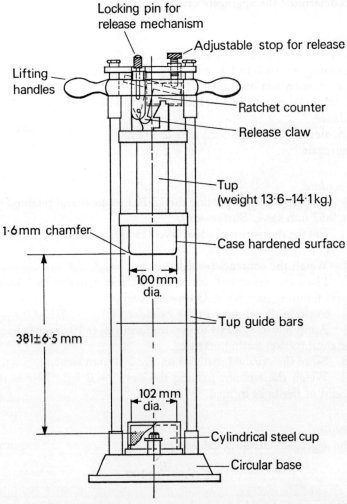

Fig 17 Aggregate impact test machine (B.S. 812)

Aggregate crushing value (A.C.V.)
The aggregate crushing value gives a relative measure of the resistance of an aggregate to crushing under a gradually applied compressive load. With aggregate of an aggregate crushing value of 30 or higher the result may be anomalous, and in such cases the 10% fines value (page 93) should be determined.

EXPERIMENT 28
To determine the aggregate crushing value.

Apparatus
Cylinder and plunger apparatus (Fig 18)
Tamping rod (450 to 600 mm long, 16 mm diam., rounded end)
Compression test machine
Metal measuring cylinder (180 mm deep, 115 mm diam.)
Balance
B.S. sieves—12·7 mm, 9·52 mm, 2·40 mm
Aggregate

Procedure
(a)   Prepare a sample passing the 12·70 mm sieve and retained on the 9·52 mm sieve. Surface-dry.
(b)   Fill the measure in 3 equal layers tamping each layer 25 times. Level off.
(c)   Weigh the contents ($=A$ kg).
(d)   Place the measured sample in the test cylinder in 3 equal layers tamping each layer 25 times. Level off.
(e)   Position the plunger in the cylinder.
(f)   Apply a load of 40·64 tonnef uniformly over 10 minutes using the compression testing machine.
(g)   Sieve the crushed material on the 2·40 mm sieve.
(h)   Weigh the amount passing the sieve ($=B$ kg). This is the weight of the fines formed.

Results
The aggregate crushing value is given by—

$$\text{A.C.V.} = \frac{B}{A} \times 100$$

The mean should be calculated from two determinations.

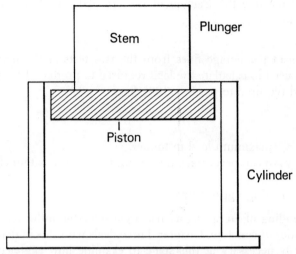

Fig 18   Cylinder and plunger apparatus for aggregate
crushing test (B.S. 812)

Ten per cent fines value
The 10% fines value gives a measure of the resistance of an aggregate to crushing which is applicable to both weak and strong aggregates.

B.S. 882 specifies that, for concrete, the 10% fines value shall be not less than 5·08 tonnef. Where the concrete is for wearing surfaces the A.I.V. shall not be less than 10·16 tonnef.

EXPERIMENT 29
To determine the 10% fines value of an aggregate.

Apparatus
As for Experiment 28.

Procedure
(a) to (f)—as for Experiment 28.
(g)   Apply the load uniformly so as to cause a total penetration of the plunger in 10 minutes of about—
15 mm—round or partly rounded aggregates
20 mm—normal crushed aggregates
24 mm—honeycombed aggregates.
(h)   Record the maximum load to produce the required penetration.

(i)   Determine the fraction of the crushed sample passing the 2·40 mm sieve.

Results

The mean percentage fines from the two tests at the maximum load is used to calculate the load required to produce 10% fines:

Load required to produce 10% fines—

$$= \frac{14x}{y+4}$$

where x = maximum load in tonnef.

y = mean percentage fines from two tests at x tonnef load.

Proportioning aggregates

The grading of an aggregate has a considerable influence on the properties of concrete. Chapter Three deals with this in detail and it is only necessary at this stage to examine how aggregates are proportioned to give a selected grading.

One method, using the *Fineness Modulus*, was illustrated on page 73. An alternative, graphical, method is given in Road Note 4 and is best explained by means of an example. Example 3 shows the application of the method to the blending of two aggregates; if it is applied to more than two then the two coarsest are combined first and the resulting grading combined with the next coarsest, and so on.

Example 3   Two aggregates are available which have the gradings given in Table 15. Proportion them to correspond with the grading curve of Fig 19.

| | Percentage passing | |
Sieve size	Coarse aggregate	Fine aggregate
19 mm	100	
10 mm	31	
5 mm	7	100
No 7	0	92
No 14		76
No 25		48
No 52		20
No 100		3

Table 15   Aggregates for Example 3

Solution

1 Draw percentage scales along three sides of the graph paper as in Fig 20.

2 Insert the grading of the fine aggregate along the left-hand vertical axis by marking points and numbering them with the sieve size so that the ordinate of each point represents the percentage of material passing that sieve.

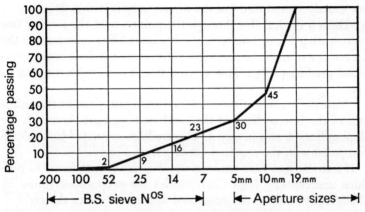

Fig 19   Grading curve for Example 3

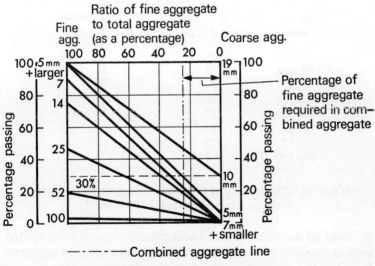

Fig 20   Graph for Example 3

3   Insert the grading of the coarse aggregate along the right-hand vertical axis in a similar manner.

4   Draw straight lines joining each point on the left-hand axis to the point with the same sieve size on the right-hand axis.

5   Having prepared the graph decide which point on the grading curve is to be accurately reproduced in the combined grading. In this case a suitable choice is that representing the percentage passing the 5 mm sieve, that is 30%. The other points on the curve are not exactly reproduced but the discrepancy is within 3%.

6   Draw a vertical line through the point where the sloping line representing 5 mm intersects the horizontal line representing the percentage of material passing a 5 mm sieve required in the combined grading (30%). This vertical line is the *combined aggregate line*.

7   Note the grading of the combined aggregate given by the ordinates of the intersections of the combined aggregate line with the sloping lines drawn in step 4. The gradings are noted in Table 16 compared with the required grading from Fig 19.

If a choice had been made in step 5 which is now found to produce unacceptably high discrepancies the proportions would be changed by shifting the combined aggregate line to the right or left to minimise them.

	Percentage passing	
Sieve size	Combined aggregate from Fig 20	Required grading from Fig 19
19 mm	100	100
10 mm	48	45
5 mm	30	30
No 7	23	23
No 14	19	16
No 25	12	9
No 52	5	2
No 100	1	0

Table 16   The grading of the combined aggregate compared with the required grading in Example 3

8   Read off the percentage of fine aggregate required using the top scale at the point where it is intersected by the combined aggregate line (25%).

Ans. The combined aggregate should consist of 25% of the fine aggregate and 75% of the coarse aggregate.

Questions

1 What are aggregates? Why are they valuable building materials?

2 Name the three classes of aggregate and give three examples of each class. Give one application of each class.

3 Name eight British Standard tests to be applied to aggregates and describe one of them in detail.

4 How does B.S. 882 describe the necessary quality of aggregates? Certain materials are said to be 'deleterious'. What are two such materials and why are they so described?

5 Outline the sampling procedure for aggregates. How is the main sample reduced to give a quantity of suitable size for testing? Illustrate your answer by a sketch of any equipment used.

6 What is meant by the term 'grading' as applied to aggregates? What is a gap-graded aggregate?

7 What is meant by (a) coarse aggregate, (b) fine aggregate, (c) all-in aggregate?

8 Describe how you would carry out the measurements which would enable you to set up a grading chart. What is the *fineness modulus* of an aggregate?

9 What are grading zones? Where would you find them tabulated?

10 Name three lightweight aggregates. What properties do they confer to plaster, mortar, and concrete?

11 Name three tests which may be applied to aggregates (a) in the field, (b) in the laboratory. What is the purpose of each test?

12 Select one test each from (a) and (b) in Question 11 and describe how they are carried out.

13 What is meant by (a) flakiness index, (b) elongation index, (c) angularity number, when applied to aggregates?

14 Why is it necessary to know the moisture content of an aggregate?

15 How is the specific gravity of an aggregate defined? Why is it necessary to take particular care in defining specific gravity in the case of aggregates?

16 What is meant by the bulking of fine aggregate?

17 What mechanical tests are carried out on aggregates? Outline one of them.

Problems

1   Columns (a), (b), and (c) in the following table give the gradings of three aggregates. Proportion them to produce the grading given in column (d).

Sieve size	Percentage passing			
	(a) *38 mm crushed rock*	(b) *19 mm crushed rock*	(c) *Sand*	(d) *Required grading*
38 mm	100			100
19 mm	14	100		50
10 mm	8	34		36
5 mm	2	6	100	24
No 7	0	0	78	18
No 14			59	12
No 25			40	7
No 52			12	3
No 100			1	0

Hint : Combine the two coarse aggregates first—34% to pass the 19 mm sieve.

Next combine these two with the sand—24% to pass the 10 mm sieve (22% of sand must be combined with a coarse aggregate containing 23% of the 19 mm crushed rock and 77% of the 38 mm crushed rock).

Five Concrete

Concrete is a rock-like material made by binding together aggregate particles with cement. It can be broadly classified into two types, dense concrete and lightweight concrete, depending on the nature of the aggregate used; there are also subdivisions within each classification.

Normal dense concrete is made by mixing water, cement, and normal aggregate (*see* Chapter Four), pouring into a mould and compacting it either by hand ramming or by mechanical vibration to remove entrapped air. High-density concrete is made by using the heavy aggregates. High-strength concrete is made from aggregates of high mechanical strength and requires a more sophisticated design procedure than the normal concrete.

Lightweight concrete is made from lightweight aggregates or by the use of air entraining agents. The properties of this material were summarised in Chapter Four.

Dense concrete

Dense concrete is characterised by high compressive strength and excellent resistance to abrasion, corrosion, and fire. The tensile strength is improved by reinforcement.

Water/cement ratio

Cement will not combine chemically with more than half the water present in the mix. Since is requires about 0·2 to 0·25 of its weight of water to fully hydrate, if the water/cement ratio is less than about 0·4 to 0·5 the hydration is incomplete. However, the strength continues to increase with a reduction of water/cement ratio to 0·2 or less. It therefore appears that only the outer surface of each cement particle is hydrated. The remaining water forms water voids or may subsequently dry out resulting in a contraction known as 'drying shrinkage'. This point will be taken up again later in this chapter. It is necessary to take care in defining the water/cement ratio which is usually written—

$$\text{water/cement ratio} = \frac{\text{weight of water in the mix}}{\text{weight of cement in the mix}}$$

In Chapter Four it was shown that the aggregate may contain water and so it follows that when the water/cement ratio is being considered it is necessary to define it in a way which takes account of this. The accepted practice is to base it on the total amount of water added to concrete made with air-dry aggregate. When the water content is different from the air-dry condition a correction is made.

The water/cement ratio is the primary factor in determining the strength and durability of ordinary concrete. As the ratio is decreased below about 0·5 however, the strength, surface texture, and shape of the aggregate begin to have a more important role and so must be taken into account, as must the aggregate/cement ratio in these regions. The influence of water/cement ratio on the compressive strength of concrete is depicted in Fig 21.

Workability
The workability, or ease of compaction, of the concrete mix will be examined with respect to the following variables:
(a)   The proportion of water in the mix.
(b)   The proportion of aggregate in the mix.
(c)   The grading of the aggregate.
(d)   The shape and texture of the aggregate.

(a)   It is readily appreciated that the water present in a concrete mix lubricates the aggregate particles enabling them to slide relative to one another. An increased water content leads to increased workability. Later in this chapter we shall deal with a concrete design method wherein it is necessary to fix the water/cement ratio for a given compressive strength. If this method is being followed then increasing the water content to increase the workability must be accompanied by a proportionate increase in cement content to maintain the water/cement ratio.
(b)   As the proportion of aggregate is increased more water is required to lubricate the particles and so maintain the workability. This again upsets the water/cement ratio with the expected reduc-

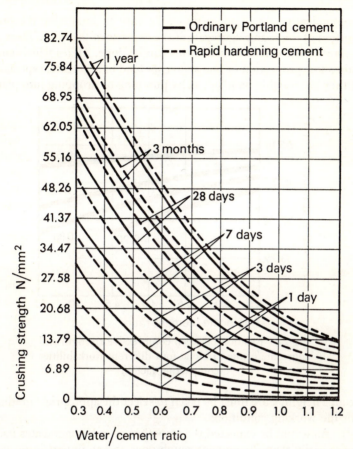

Fig 21 Relation between compressive strength and water/cement
ratio for 100 mm cubes of fully compacted concrete for mixes of
various proportions (adapted from Road Note 4)

tion in compressive strength. The relation between the aggregate/
cement ratio and strength as the workability is varied is shown in
Fig 22.

(c) The larger sizes of aggregate require less water for a given
workability than do the finer sizes and their effect is more pro-
nounced with mixes of high aggregate/cement ratio; rich mixes
are slightly affected. As the water/cement ratio is increased the
proportion of coarse aggregate required to provide a given workability decreases. For a given water/cement ratio it seems more

economical to obtain adequate workability by increasing the proportion of the coarse aggregate at the expense of the sand. There is
a limit to the extent to which this can be done. Mixes which contain a low proportion of fine aggregate tend to segregate, especially
if they are roughly handled, as by pumping or dropping into place

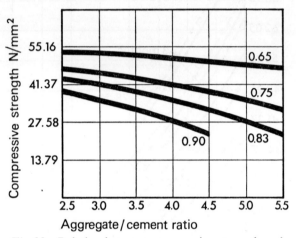

Fig 22   Relation between compressive strength and
aggregate/cement ratio at four different workabilities
(compacting factor)

from a height. Care must be taken to select the appropriate grading
for the working conditions.
(d)   As would be expected the smoother rounder aggregates pack
more easily than the angular rough ones and so impart improved
workability to the mix.

When considering the maximum size of aggregate for a particular purpose it is common practice to choose a size not greater
than one quarter of the minimum thickness of the cured concrete.
This means that 10 mm aggregate will be used for small precast
work, 19 mm for reinforced and prestressed concrete generally,
while the 38 mm aggregate is reserved for roads and mass concrete
structures. The still larger sizes are used in such structures as
dams.

As was mentioned in Chapter Four fine dust, silt, and clay are
unsuitable for concrete since they coat the surface of the aggregate
and interfere with the bond between the latter and the cement
paste.

Shrinkage of concrete and moisture movement

Two stages of concrete shrinkage can be distinguished.

1 *Plastic shrinkage*

Plastic shrinkage is an initial shrinkage occurring before the paste sets, while it is still in the plastic state. It is due to loss of water by seepage through the formwork, by absorption by the formwork, and by evaporation. It is offset to some extent by the thermal expansion caused by the exothermal nature of the reaction between the cement and the water. In neat cement paste the shrinkage may be about 0·6% in the first few hours but in concrete it is generally about 0·1 to 0·2%.

The extent of the shrinkage can be minimised by taking precautions during the curing: prevention of seepage and absorption by careful erection of formwork and the correct choice of materials and the control of evaporation by covering the surface with tenting or plastic sheeting.

Plastic shrinkage is unlikely to cause cracking.

2 *Drying shrinkage*

Drying shrinkage occurs as the concrete hardens and is allowed to dry. It is due to the shrinkage of the cement gel and is not fully reversible although on subsequent wetting it will swell but not to such an extent as to recover its initial volume. The irreversible part of the shrinkage is usually about 0·3% of the drying shrinkage. If the drying and wetting is continued the shrinkage and swelling become practically reversible. These changes are known as 'moisture movements' and, along with the drying shrinkage, result in stresses being produced within the concrete which in turn give rise to cracking. The stresses may be relieved by *creep* in plain concrete or by the incorporation of reinforcement. They can also be relieved by suitably placed joints; floors, long walls, and roofs are cast in panels to reduce cracking. This procedure also reduces cracking due to thermal movements. It will be appreciated that the stresses are increased if there is a considerable difference in the drying rate at different points in the structure as there will be between, say, the foundations and an outside wall exposed to sun and wind. The stresses are reduced if the concrete is allowed to dry out slowly, not only because of the increased strength arising from the more extensive reaction between the water and the cement but also because of the increased duration of the plastic flow.

Finally, drying shrinkage and moisture movements are increased by the use of:

(a) Rapid hardening cements.
(b) Rich mixes.
(c) High water/cement ratios.
(d) Finer aggregates.
(e) Porous aggregates.
(f) Flint gravel aggregates instead of limestone or igneous rock aggregates.

Mix design

The purpose of mix design is to produce a concrete having the desirable properties such as workability, strength, and durability in the most economical way. In order to achieve this it is necessary to consider simultaneously the properties of both the fresh and hardened concrete:

(a) Fresh concrete. Ensure that the workability and resistance to segregation are suitable for the working conditions, the mixing, transporting, and placing.

(b) Hardened concrete. Ensure that the strength, weathering, and abrasion resistance, water impermeability, and appearance are suitable for the intended purpose.

Mix design is generally based on considerations of compressive strength since it is easily measured and has been found to be a guide to the general quality of the concrete. The other desirable properties are usually associated with a high compressive strength. Experience shows that for properly compacted medium strength concrete (strength not exceeding $4 \cdot 0 \ N/mm^2$) the compressive strength is mainly dependent on the water/cement ratio. Thus once the water/cement ratio has been decided it is only necessary to select the aggregate/cement ratio (richness) and the fine aggregate/coarse aggregate ratio to give the required workability and cohesiveness.

The procedure for high strength concrete is more complex being dependent on the type and proportion of aggregate used. Different procedures are also followed for *dry lean* and lightweight concretes.

Design of medium strength concrete

The design procedure for ordinary concrete is as follows.

(a) *Average strength*

Decide the average strength required. This is calculated from the minimum strength and makes allowance for the normal variation of test cubes prepared under works conditions; the mix is designed to have a mean strength that exceeds the specified works cube strength by twice the calculated standard deviation. This value is chosen because it provides an acceptable control margin; examination of the statistical Normal Distribution curve shows that only 2·5% of the results may be expected to fall below the mean value by more than two standard deviations. The standard deviation should be calculated from at least 40 individual works test cubes each representing separate batches of similar concrete produced by the same plant and under the same supervision.

It may be calculated from the formula—

$$\text{Standard deviation} = \sqrt{\left\{ \frac{(x - \bar{x})^2}{N - 1} \right\}}$$

where N = the number of results.
x = the individual result.
$\bar{x}$ = the average of the results.

If a calculating machine is available the above formula is suitably rearranged.

British Standard Code of Practice 114:Part 2:1969 recommends that mixes should be designed to satisfy any specified works cube strength requirement within the range given in Table 17. The mix limitations given in the table should be observed.

C.P. 114 further recommends that no standard deviation less than 3·5 N/mm² should be used as a basis for designing a mix and that, in the absence of previous information, a standard deviation of 7 N/mm², i.e. a margin of 14 N/mm², should be used initially. This figure will be adjusted to accord with the results of the works cube tests, as mentioned previously, when sufficient are available.

(b) *Water/cement ratio*

Having decided the average strength required select the water/ cement ratio to suit (Fig. 21). This is the lesser of that needed to give the necessary average strength and the required durability; in

(1)	(2)
Mix limitations	*Specified works cube strength (N/mm²)*
For Portland cement or Portland blast-furnace cement concrete, not less than 240 kg nor more than 540 kg of cement per cubic metre of finished concrete; and designed for the required concrete strength at 28 days, within the range of Column 2.	15–50 at 28 days
For high-alumina cement concrete, not less than 270 kg nor more than 420 kg of cement per cubic metre of concrete; and designed for the required concrete strength at one day, within the range given in Column 2, with a water/cement ratio not more than 0·50.	

Table 17   Strength requirements for designed concrete mixes
(C.P. 114)

some situations it may be necessary to use a richer mix to provide adequate durability (*see* Table 18).

(c)   *Workability*
Decide the degree of workability required for the conditions of placing (Table 22). In general when the compacting factor is less than 0·8 compaction by vibration is necessary and for values less than 0·7 the vibration must be assisted by pressure on the top of the concrete.

(d)   *Aggregate/cement ratio*
Determine the aggregate/cement ratio required to give the selected workability at the fixed water/cement ratio using tables such as those produced by the Cement and Concrete Association and the Road Research Laboratory (Road Note 4). One such set of data is reproduced in Table 18 and should be used in conjunction with Fig 23.

(e)   *Grading of aggregates*
Proportion the aggregates to give the selected grading. One method of doing this was described in Chapter Four (p. 44).

(f)   *Trial mixes*
Prepare trial mixes, test in accordance with B.S. 1881:1970, and

Degree of durability	Very Low				Low				Medium				High			
Grading of aggregate (Curve No on Fig 3)	1	2	3	4	1	2	3	4	1	2	3	4	1	2	3	4
Water/cement ratio by weight 0·35	3·7	3·7	3·5	3·0	3·0	3·0	3·0	2·7	2·6	2·6	2·7	2·4	2·4	2·5	2·5	2·2
0·40	4·8	4·7	4·7	4·0	3·9	3·9	3·8	3·5	3·3	3·4	3·5	3·2	3·1	3·2	3·2	2·9
0·45	6·0	5·8	5·7	5·0	4·8	4·8	4·6	4·3	4·0	4·1	4·2	3·9	×	3·9	3·9	3·5
0·50	7·2	6·8	6·5	5·9	5·5	5·5	5·4	5·0	4·6	4·8	4·8	4·5	×	4·4	4·4	4·1
0·55	8·3	7·8	7·3	6·7	6·7	6·2	6·0	5·7	×	5·4	5·4	5·1	×	4·8	4·9	4·7
0·60	9·4	8·6	8·0	7·4	6·8	6·9	6·7	6·2	×	6·0	6·0	5·6	×	×	5·4	5·2
0·65	—	—	—	8·0	7·4	7·5	7·3	6·8	×	×	6·4	6·1	×	×	5·8	5·6
0·70	—	—	—	—	8·0	8·0	7·7	7·4	×	×	6·8	6·6	×	×	6·2	6·1
0·75	—	—	—	—	—	—	—	7·9	×	×	7·2	7·0	×	×	6·6	6·5
0·80	—	—	—	—	—	—	—	—	×	×	7·5	7·4	×	×	×	7·0
0·85	—	—	—	—	—	—	—	—	×	×	7·8	7·8	×	×	×	7·4
0·90	—	—	—	—	—	—	—	—	×	×	×	8·1	×	×	×	7·7

Notes   (1) These proportions are based on specific gravities of approx. 2·5 for the coarse aggregate and 2·6 for the fine aggregate.

(2) × indicates that the mix would segregate.

(3) — indicates that the mix was outside the range tested.

Table 18   Aggregate/cement ratio required to give four different degrees of workability with different gradings of irregular aggregate (max. size 19 mm) (by permission of the Road Research Laboratory—Road Note 4)

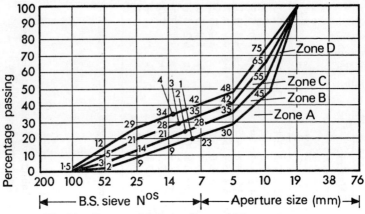

Fig 23   Curves of four gradings of 19 mm aggregate
(from Road Note 4)

adjust the trial mix proportions if necessary. Re-check in accordance with B.S. 1881.

B.S.C.P. 114 stipulates that representative samples of the materials to be used should be mixed using the proposed proportions on each of 3 different days. The workability of each of these 3 trial mixes should be determined, and a batch of 6 cubes from each should be made, 3 for test at 7 days and 3 for test at 28 days. The mix is acceptable if the average strength of the 3 trial mixes is not less than the specified works cube strength plus the designed standard deviation. It is recommended that further trial mixes be made if the range of the 3 cube results in any batch exceeds 15% of the average of that batch, or if the range of the 3 batch averages exceeds 20% of the overall average of the batches.

The trial mixes will allow adjustments to be made not only with respect to the strength but also with respect to workability, cohesiveness, and surface finish. These adjustments will be made by altering the aggregate/cement ratio, the fine aggregate/coarse aggregate ratio or both, keeping the water/cement ratio in line with the strength requirement.

Concrete incorporating high alumina cement requires only 3 cubes from each trial 6. These will be tested for 1 day strength.

(g) *Works cube tests*
Prepare a full-scale works mix for final checking using the works plant.

B.S.C.P. 114 stipulates that when a mix is used for the first time the level of control and the suitability of the mix proportions should be established by obtaining a large number of results as soon as possible.

The procedure to be adopted for obtaining these results is as follows:

1 Take a random sample on 8 separate occasions during each of the first 5 days of using the mix.

2 Take at least 1 sample thereafter on each day the particular mix is made. The number of samples and the times of sampling should be varied at random.

3 Prepare 2 cubes from each sample, one for a 7 day test and the other for a 28 day test.

A statistical check is then made in the following way:

(a) Examine the works cube results both individually and in non-overlapping consecutive sets of 4. Calculate the average and range of each set.

(b) Modify the mix proportions if in the first ten consecutive sets any of the undermentioned conditions are not satisfied:

(i) Not more than 2 individual results of the 40 cubes should fall below the specified works cube strength.

(ii) No value of the range in any set should exceed 4 times the designed standard deviation.

(iii) Not more than one set should have an average which is less than the specified strength plus 4/3 times the designed standard deviation.

(iv) No value of the average for any set should be less than the specified strength plus the designed standard deviation.

(c) After 10 consecutive sets of results have been obtained calculate the overall average and the standard deviation of the 40 results and modify the mix as appropriate.

(d) Subsequently, if any of the foregoing conditions are not satisfied, calculate the overall average and the standard deviation of the previous consecutive 40 results. If the overall average strength minus twice the standard deviation is less than the specified works cube strength, modify the mix as appropriate.

Where a high alumina cement mix is being used only 1 cube need be made from each sample of concrete. The control is then based on the results of tests on the cubes at an age of 1 day.

Example 4 Design a concrete mix having a 28 day works cube

strength of 28 N/mm². The degree of control is 'very good' allowing a standard deviation of 3·5 N/mm² to be taken. Ordinary
Portland cement is to be used and the aggregates available have the
grading given in Example 3 (p. 94). The aggregates are air-dry.
Low workability is required.
(a)   Design cube strength $= 28 + (2 \times 3·5) = 35$ N/mm².
(b)   Water/cement ratio.
From Fig 21 a cube strength of 35 N/mm² requires a water/cement
ratio $= 0·53$ (28 days).
(c)   Workability.
The workability is required to be *low*.
(d)   Aggregate/cement ratio.
Reference to Table 18 indicates that at a water/cement ratio of
0·53 and low workability the aggregate/cement ratio will need to
be, by interpolation for an aggregate grading curve 1 or 2 (Fig 23)—

$$(6·2 - 5·5) \times \tfrac{3}{5} + 5·5 = \frac{0·7 \times 3}{5} + 5·5$$
$$= 0·4 + 5·5$$
$$= 5·9$$

(e)   Grading of aggregates.
The aggregates available have the grading of Example 3 (p. 94).
This example showed that if they were combined using 25% fine
aggregate, that is using a fine aggregate/coarse aggregate ratio of
1 :3, then their combined grading closely corresponded to curve 1
in Fig 2. With the usual notation this is:
    100/48/30/23/19/12/5/1 compared with 100/45/30/23/16/9/2/0 of
    curve 1.
(f)   Quantities of materials required for a 1000 kg batch.

$$\text{Cement} = \frac{1000}{1 + 0·53 + 5·5} = 142·2 \text{ kg}$$

Aggregate $= 142·2 \times 5·5 \quad = 782·3$ kg (combined)
                              (fine/coarse $= 1:3$)
    Water $= 142·2 \times 0·53 = 75·4$ kg
Note : The aggregates were supplied 'air-dry'. If they were otherwise the added water would need adjustment with a corresponding
adjustment in the weight of the aggregates.

Nominal mixes

Nominal mixes are the traditional way of proportioning concrete materials. The proportions are expressed by volume. Thus a 1:2:4 specification means 1 part of cement:2 parts of fine aggregate:4 parts of coarse aggregate, by volume. In the case of fine aggregate it is necessary to allow for *bulking* which occurs if it is damp (*see* Chapter Four). In practice this usually means increasing the specified volume of sand by about 25%. Tables 19 and 20 give the nominal mixes allowed by B.S.C.P. 114. Although the mixes are *specified* by volume the actual *batching* is preferably carried out by weight, requiring that the densities of the materials be known.

Example 5   Calculate the quantities of materials required for a 1:2:4 nominal mix based on 100 kg of cement if the batching is to be carried out by weight. Assume that the bulk densities are as follows:

Cement $= 1440$ kg/m³
Fine aggregate $= 1640$ kg/m³
Coarse aggregate $= 1390$ kg/m³
100 kg cement $= 100/1440$ m³ $= 1/14 \cdot 4$ m³
Hence 1 part by volume of cement $= 1/14 \cdot 4$ m³
    2 parts by volume fine aggregate $= (2 \times 1640)/14 \cdot 4$
                         $= 227 \cdot 8$ kg
    4 parts by volume coarse aggregate $= (4 \times 1390)/14 \cdot 4$
                         $= 386 \cdot 1$ kg

Ans.   The quantities of materials required are:
             100 kg cement
             228 kg fine aggregate
             386 kg coarse aggregate.

The above example assumes that the materials are dry; if they are damp the weights must be adjusted to compensate for the water contained in each material.

The proportions of cement to fine aggregate and coarse aggregate should be as given in the tables except that for Portland cement and Portland blast-furnace cement intermediate proportions may be used provided that the ratio of 1:2 is maintained for the fine to coarse aggregate volumes.

If a denser or more workable concrete can be produced by a variation in the ratio of the fine aggregate to coarse aggregate it may be varied within the limits of $1:1\frac{1}{2}$ and 1:3. The sum of the volumes of the aggregates must however remain unaltered appropriate to the nominal mix (or the intermediate mix where applicable).

Mix proportions	Cubic metres of aggregate per 50 kg cement		Cube strength within 28 days after mixing (N/mm²)		Alternative cube strength within 7 days after mixing (N/mm²)	
	Fine	Coarse	Prelim. Test	Works Test	Prelim. Test	Works Test
1:1:2	0·035	0·07	40	30	26·7	20
1:1½:3	0·05	0·10	34	25·5	22·7	17
1:2:4	0·07	0·14	28	21	18·7	14

Table 19 Proportions and strength requirements for nominal concrete mixes with Portland cement or Portland blast-furnace cement and with aggregates complying with B.S. 882 or B.S. 1047 (C.P. 114)

Mix proportions	Cubic metres of aggregate per 50 kg of cement		Cube strength within 1 day after mixing (N/mm²)	
	Fine	Coarse	Preliminary test	Works test
1:2:4	0·07	0·14	46·7	40

Table 20 Proportions and strength requirements for a nominal concrete mix with high alumina cement and with aggregate complying with B.S. 882 or B.S. 1047 (C.P. 114)

The ratio may also be adjusted to suit the grading of the fine aggregate and the type and maximum size of the coarse aggregate, the richness and workability of the mix, and the required resistance to segregation as dictated by the conditions of mixing, transporting, and placing. For example the following ratios may be suitable for a Zone 2 sand with the aggregate of the maximum size stated:

Maximum size of aggregate	10 mm	19 mm	38 mm
Ratio (fine:coarse)	1:1½	1:2	1:3

The ratio should, however, be increased as the fineness of the sand increases. Thus, using a 19 mm coarse aggregate, the most suitable proportions may be:

Zone of fine aggregate	1	2	3
Ratio (fine:coarse)	1:1½	1:2	1:3
Complete mix ratio (cement:fine:coarse) (approx.)	1:2½:3½	1:2:4	1:1½:4½

Zone 4 sand is not considered suitable for use with nominal mixes for structural work, whatever coarse aggregate size is adopted, but it can be used for special mixes.

If it is necessary to use *all-in* aggregate instead of fine and coarse aggregate separately the mix proportions corresponding to 1:2:4 would be 1:5 and not 1:6 as expected. This is because the fine aggregate tends to fill the voids in the coarse aggregate and the total volume is less than the sum of the volumes of the fine and coarse aggregate measured separately.

If a coarse aggregate smaller than the usual 19 mm is used the workability and compressive strength is maintained by using a richer mix. For example a 1:2:3 mix may be used instead of a 1:2:4 mix. Likewise a richer mix would be required if a rougher or more angular aggregate were used.

In choosing the water/cement ratio the quantity of water used should be just sufficient to produce a dense concrete of adequate workability for its purpose.

The present practice is to use nominal mixes with lightweight aggregates where strength is not important.

This method of specifying mixes has the advantage of simplicity and is easily remembered, but it disregards other variables which affect the strength and durability of the final product. Apart from the richness of the mix the following variables should be taken into account:

(a)   The nature of the materials, in particular their quality.
(b)   The conditions of mixing.
(c)   The conditions of placing.
(d)   The degree of quality control which can be exercised over the final product.

A full mix design procedure is necessary to cover all these variables but in certain cases standard mixes are permitted.

Standard mixes

Nominal mixes are being replaced in structural work by standard mixes in which the proportions of materials are specified by weight and are based on the major factors known from experience of mix design to affect the minimum compressive strength of concrete. The standard mixes are recommended where a full mix design procedure is impracticable.

Table 21 sets out the standard mixes given in C.P. 114 to

E

produce three grades of concrete. Since these proportions have been shown to be satisfactory for producing concrete of the three specified strengths indicated, assuming a standard deviation of 7 N/mm², the trial mixes necessary for the full mix design procedure can be dispensed with. When the works cube test results are unsatisfactory the mix should be altered by using the proportions for the next higher strength of concrete. Where there is no higher strength included in the table the cement content should be increased by 10%. The revised mix should continue to be used until 40 works cube test results have been obtained and the mean and the standard deviation of these results calculated. If the strength is unnecessarily high, further modifications to the mix can be made having regard to the data from the works cubes of both the original and modified mixes.

Standard mixes are also allowed by C.P. 116 'The structural use of precast concrete' where, in addition to the same mixes as given in C.P. 114, data are given for application where the standard deviation does not exceed 3·5 N/mm², that is where the degree of control is very good.

The following points should be noted in connection with Table 21:

(a)	The mixes only apply to concrete made with cement complying with B.S. 12 (Portland cement, ordinary or rapid-hardening) or B.S. 146 (Portland blast-furnace cement).

(b)	Aggregates should comply with B.S. 882 (Concrete aggregates from natural sources) or B.S. 1047 (Air-cooled blast-furnace slag coarse aggregate for concrete).

(c)	Fine aggregate in Grading Zone of B.S. 882 and air-entraining agents should not be used.

(d)	If aggregate complying with B.S. 1047 or other lightweight aggregates are used the producer's recommendations should be followed.

(e)	The weights given are based on a fine aggregate having a grading within the limits of Grading Zone 2 in B.S. 882. If Zone 1 aggregate is used its weight should be *increased* by at least 10 kg. If Zone 3 aggregate is used its weight should be *decreased* by at least 10 kg. The weight of coarse aggregate should be adjusted correspondingly to maintain the same total weight of aggregate. Larger adjustments are more likely with the leaner mixes.

(f)	If a crushed stone sand or a crushed gravel sand is used instead of natural sand the weight of the coarse aggregate should

Specified works cube strength at 28 days N/mm²	Weight of dry sand per 50 kg of cement kg	Weight of coarse aggregate per sq kg of cement											
		10 mm maximum size			13 mm maximum size			19 mm maximum size			38 mm maximum size		
Workability Slump (mm) compacting factor		Low 0–5 0·80–0·86	Medium 5–25 0·86–0·92	High 25–50 0·92–0·97	Low 5–20 0·81–0·87	Medium 20–40 0·87–0·93	High 40–100 0·93–0·97	Low 12–25 0·82–0·88	Medium 25–50 0·88–0·94	High 50–125 0·94–0·97	Low 25–50 0·82–0·88	Medium 50–100 0·88–0·94	High 100–175 0·94–0·97
	kg	kg	kg	kg	kg	kg	kg	kg	kg	kg	kg	kg	kg
21	90	145	110	90	165	135	110	190	155	135	225	190	165
25·5	80	125	90	65	145	110	99	165	135	110	200	165	145
30	65	100	—	—	125	90	—	145	110	90	165	135	110

Table 21   Standard mixes (C.P. 114)

be reduced by at least 10 kg without altering the weight of sand.
(g)   If single-sized aggregates are used they should be proportioned to produce combined gradings within the limits of B.S. 882 or B.S. 1047 for graded aggregate of the appropriate size.
(h)   If the specific gravity of either the coarse or fine aggregate differs significantly from 2·6 the weight of each type of aggregate should be adjusted in proportion to the specific gravity of the materials.
(i)   In selecting the mix proportions due regard should be paid to the durability requirements which may indicate a richer mix. For example, a concrete having a specified works cube strength of 25 N/mm² at 28 days may be suitable structurally and when used in exposed conditions in a non-industrial area; however if the structures were sited in an industrial area, a higher grade of concrete with a greater strength would probably be required.

Concrete testing

British Standard 1881:1970 'Methods of testing concrete' applies to concrete whose nominal maximum aggregate size does not exceed 40 mm. It is published in five parts.

Part 1   *Methods of sampling fresh concrete.*

Part 2   *Methods of testing fresh concrete.*
(a)   Slump test.
(b)   Compacting factor test.
(c)   'V-B' consistometer test.
(d)   Determination of weight per cubic metre of fresh concrete.
(e)   Determination of air content of fresh concrete.
(f)   Analysis of fresh concrete.

Part 3   *Methods of making and curing test specimens.*
(a)   Making and curing test cubes.
(b)   Making and curing no-fines test cubes.
(c)   Making and curing test beams.
(d)   Making and curing test cylinders.

Part 4   *Methods of testing concrete for strength.*
(a)   Test for compressive strength of test cubes.
(b)   Preparation and compression testing of drilled cores from concrete.

(c)   Preparation of sawn beams.
(d)   Test for flexural strength.
(e)   Test for compressive strength using portions of beams broken in flexure ('equivalent cube' method).
(f)   Test for indirect tensile strength of cylinders.

Part 5   *Methods of testing hardened concrete for other than strength.*
(a)   Determination of the saturated and dry densities of concrete specimens by the water displacement method.
(b)   Test for the static modulus of elasticity by means of an extensometer.
(c)   Test for the dynamic modulus of elasticity by an electro-dynamic method.
(d)   Determination of changes in length on drying and wetting (initial drying shrinkage, drying shrinkage, and wetting expansion).
(e)   Test for determining the initial surface absorption of concrete.
(f)   Test for water absorption.

The slump test
The slump test was described in Chapter Three. It is a test commonly carried out in the field but is only applicable to mixes of high workability. Stiffer mixes, which are required when mechanical vibration is employed, should be tested by the compacting factor test of the 'V-B' consistometer test.

The compacting factor test
Intended originally as a laboratory test the compacting factor test is being increasingly used in the field and is the only one sufficiently sensitive for mixes of low workability. If strictly comparable results are to be obtained each test should be carried out at a constant time interval after the mixing is completed. This is because the sensitivity is high enough to detect differences in workability arising from the initial processes in the hydration of the cement. It is recommended that the test be carried out two minutes after mixing. If the test is to be made on concrete which is not fresh it should be remixed without any correction of water content and tested two minutes later.

   The test measures the degree of compaction of concrete which has been caused to fall through a standard height. The compacting factor is defined as the ratio of the weight of partially compacted concrete to the weight of fully compacted concrete.

EXPERIMENT 30
To measure the workability of concrete by the compacting factor test.

Apparatus
Compacting factor apparatus (Fig 24)
Plasterer's trowels (2)
Hand scoop (150 mm long)
Tamping rod (300 mm long, 15 mm diameter, rounded end)
Balance (to weigh 20 kg to the nearest 10 g)
Cement
Fine aggregate
Coarse aggregate

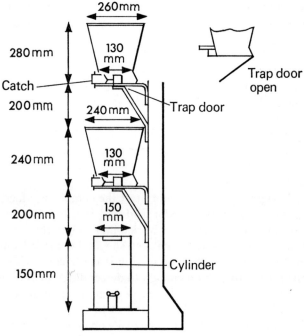

Fig 24   Compacting factor apparatus

Procedure
(a)   Prepare six separate dry 1:2:4 mixes using:
  cement—2·5 kg

fine aggregate—5·0 kg

coarse aggregate—10·0 kg (not greater than 38 mm).

(b)   To the first mix add sufficient water to produce a water/cement ratio = 0·5, that is 1·25 kg.

(c)   Mix thoroughly to a uniform consistency.

(d)   Secure all the trapdoors in the apparatus and cover the cylinder.

(e)   Place the concrete gently in the upper hopper using the hand scoop.

(f)   Make the surface of the concrete level with the brim of the hopper.

(g)   Open the top trapdoor so that the concrete falls into the lower hopper.

(h)   Immediately the concrete has come to rest *uncover the cylinder* and open the lower trapdoor.

(i)   Remove the excess of concrete remaining above the level of the top of the cylinder. This is done by holding a trowel in each hand, with the plane of the blades horizontal, and moving them simultaneously, one from each side, across the top of the cylinder. Keep the trowels pressed on the top edge of the cylinder during this operation.

(j)   Weigh the concrete contained in the cylinder. This is the 'weight of partially compacted concrete'.

(k)   Refill the cylinder from the same sample of concrete in layers approximately 50 mm deep, heavily ramming each layer so as to obtain full compaction. Strike off level.

(l)   Weigh the concrete contained in the cylinder. This is the 'weight of fully compacted concrete'.

(m)   Repeat the procedure (c) to (l) with each of the remaining mixes using the following water/cement ratios: 0·40, 0·45, 0·60, 0·70, 0·80.

Conclusion

(a)   Calculate the compacting factor of each sample of concrete—

$$\text{Compacting factor} = \frac{\text{Weight of partially compacted concrete}}{\text{Weight of fully compacted concrete}}$$

(b)   Tabulate the results and represent them graphically, plotting water/cement ratio against compacting factor (x-axis).

(c)   Using Table 22 state the use and placing condition for which each concrete may be suitable.

Degree of workability	Appox. slump (mm)	Compacting factor	Maximum aggregate size (mm)	Suitable use and placing conditions
'Very low'	0	0·75	10·0	Roads and simple rein-
	0–15	0·78	19	forced sections compacted
	0–25	0·78	38	by power operated
				machines
'Low'	0–5	0·83	10·0	Roads vibrated by hand
	15–25	0·85	19	operated machines,
	25–30	0·85	38	simple reinforced sections
				vibrated by hand operated
				machines, and mass con-
				crete without vibration
'Medium'	5–25	0·90	10·0	Hand compaction in
	25–50	0·92	19	simple reinforced sections,
	50–100	0·92	38	roads, and slabs. Heavily
				reinforced sections with
				vibration
'High'	25–100	0·95	10·0	Not normally suitable for
	50–125	0·95	19	vibration. For hand com-
	100–175	0·95	38	paction of heavily rein-
				forced or complicated
				sections

Table 22   Suggested uses and placing conditions for concrete
of different degrees of workability

The V-B consistometer test

The V-B consistometer test is particularly useful for concrete mixes of low workability and for air-entrained concrete. The apparatus is shown in Fig 25.

The main parts of the consistometer are a metal cylindrical container (A) which can be clamped to the top of the vibrating table (G), a conical mould, a funnel (D), and a transparent horizontal disc (C) mounted on a swivel arm (N).

The procedure may be summarised as follows:

(a)   Secure the container firmly to the table by means of the wing nuts (H).

(b)   Place the mould concentrically in the container; tighten F.

(c)   Swing the funnel into position.

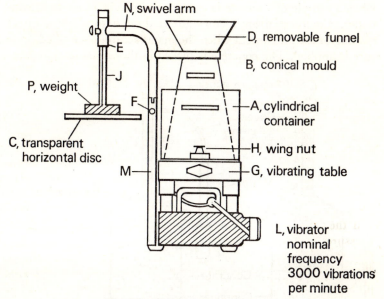

Fig 25   The V-B consistometer

(d)   Fill the mould in four layers tamping each 25 times.
(e)   Swing the funnel away and strike the mould level.
(f)   Remove the mould carefully, using a vertical lifting action.
(g)   Swing the disc (C) into the container and lower carefully until it touches the concrete. Measure the slump, using scale (J).
(h)   Tighten screw (F).
(i)   Start the vibrator and the stop-watch simultaneously.
(j)   Note the time taken for the transparent disc to become covered with cement grout. This is the time for full compaction.
(k)   Record the workability of the concrete as 'T' V-B seconds to the nearest 0·5 V-B's. Record the slump obtained in (g).

Weight per cubic metre of fresh concrete
A calibrated container holding the fully compacted concrete is weighed and the weight of the concrete determined.

Air content of fresh concrete
The air content is determined using an apparatus similar to that shown in Fig 26.
      The air content is measured by noting the reduction in volume

when a sample of concrete is compressed inside an air-tight container. Commercial air meters are calibrated to allow a direct reading of the air content when the manufacturer's specified pressure is applied. The concrete is placed in the container and com-

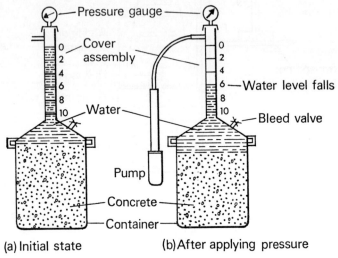

(a) Initial state   (b) After applying pressure

Fig 26   The determination of the air content of fresh concrete

pacted by vibration. After levelling off the cover assembly is fitted and the tube filled with water to the zero mark. The specified pressure is then applied and the air content is read off from the new water level. The pressure should then be released allowing the water to rise. The difference, if any, from zero of the final water position is due to absorption by the aggregate and is subtracted from the air content obtained previously. Caution should be exercised in the application of this test to lightweight aggregates.

Test for the compressive strength of test cubes
Concrete cubes are prepared for this test by curing the concrete in a mould (Fig 27) under standard conditions. The cubes are usually 150 mm size but if the nominal maximum aggregate size does not exceed 25 mm then 100 mm cubes may be used. After curing the cubes are crushed in a compression testing machine. Cubes for testing at 24 hours are tested in the moist condition with the exception of no-fines which are prepared and cured in a

different manner from other types: they are tested in the air-dry condition.

The following experiment may be modified to suit class requirements. If there are sufficient groups to give a minimum of twenty

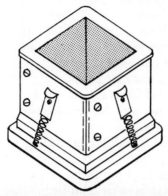

Fig 27   Steel mould for test cubes

results they may all use the same water/cement ratio enabling the standard deviation of the results to be calculated. Alternatively if the groups use different water/cement ratios the relation between water/cement ratio and strength may be obtained.

EXPERIMENT 31
To prepare test cubes and test for compressive strength.

Apparatus
Cube moulds (150 mm) (6) and mould oil
Tamping bar (380 mm long, 25 mm square ramming face, weight 1·8 kg)
Trowels (1 each for gauging and levelling)
Matting or sackcloth
Polythene sheet
Curing tank
Mixing trays
Measuring cylinder (500 ml)
Balance
Compression testing machine
Cement
Coarse aggregate (maximum size 19 mm)
Fine aggregate

Procedure
(a)   Prepare two mixes, using the following weights (kg).
    Each mix will fill three moulds:

	i	ii
Cement	8·0	5·0
Coarse	16·0	20·0
Fine	8·0	10·0

Use sufficient water in each case to provide a workable mix and note the amount of water used.
(b)   Oil the moulds and base plates.
(c)   Fill the moulds in layers approximately 50 mm deep compacting each layer 35 times with the tamping bar. Strike off level.
(d)   Label the moulds.
(e)   Cover the moulds with damp matting and then with polythene sheet. Allow to stand for 24 hours.
(f)   After 24 hours label the specimens, remove them from the moulds, and immerse in water until air bubbles cease to rise.
(g)   Place the specimens in polythene bags for 5 days.
(h)   Allow to air-dry until they are 7 days old from placing in the moulds.
(i)   Test in compression testing machine applying the load at approximately 15 N/mm² per minute.

Results
Calculate the compressive strength of each cube by dividing the maximum load by the nominal cross-sectional area; express the result to the nearest 0·5 N/mm².

Conclusion
(a)   Obtain the results from other groups in the class and calculate the standard deviation of the strength of each mix composition. The results used must all have the same water/cement ratio.
(b)   If the groups have used different water/cement ratios plot graphs relating water/cement ratio and strength.

Test for flexural strength
The flexural strength test is carried out on beams prepared by casting in steel moulds. The standard size is 150 mm × 150 mm × 750 mm long but beams 100 mm × 100 mm × 500 mm long may be used if the nominal maximum aggregate size does not exceed 25 mm. The beams are cured in a humid atmosphere in a manner

similar to that applied to the cubes for compressive strength. They are not allowed to dry before testing.

The test machine has two pairs of steel rollers having a nominal diameter of 38 to 40 mm and a length at least 10 mm greater than the width of the beam. One pair of rollers is used to support the beam and the other pair to apply the load (Fig 28).

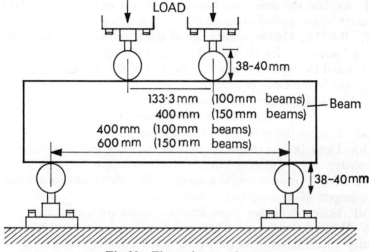

Fig 28   Flexural strength test

The load is applied at a steady rate of 4000 N/min (100 mm beams) or 9000 N/min (150 mm beams) until the specimen fails. The flexural strength is expressed as the modulus of rupture in N/mm².

Questions
1 Outline the steps to be taken in the design of structural concrete.
2 Write notes on the interrelationship of the following in the production of good quality concrete:
(a)   Water/cement ratio.
(b)   Aggregate shape and texture.
(c)   Aggregate grading.
(d)   Aggregate/cement ratio.
3 Draw graphs to illustrate the effect on the compressive strength of concrete of varying:

(a)   The water/cement ratio.

(b)   The aggregate/cement ratio.

State which of these is critical in determining the following properties of concrete. How is the particular property affected by an increase of the ratio concerned?

(a)   Workability.

(b)   Durability.

4   Explain the difference between nominal mixes and standard mixes when applied to concrete. When are they used?

5.   B.S.C.P. 114 'Structural use of reinforced concrete in buildings' advocates for the normal range of strength and workability required in reinforced concrete the use of the standard mixes as quoted in the Code of Practice.

A typical 'standard mix' quotation is

C.P. 114   25·5   13 mm   Low.

(a)   Explain the meaning of this quotation.

(b)   Describe one control method of assessing the workability of concrete.

(c)   Why is it necessary to determine the zone number of the fine aggregate when designing a concrete mix?

(d)   State three cases where standard mixes are not suitable.

6   What shrinkages are likely to occur in concrete and how do they arise? What is meant by the term 'moisture movement' as applied to concrete? What undesirable effect do these have on concrete and how may it be minimised?

List the factors influencing shrinkage and moisture movement in concrete.

7   List the tests given in B.S. 1881:1970 applicable to fresh concrete and describe one of them in detail.

8   List the tests given in B.S. 1881:1970 applicable to hardened concrete and describe one of them in detail.

Six Building Stone

Classification

Natural building stones are classified according to the way in which their parent rock has been formed. Within each class there are variations of mineral content, both in proportion and type, and it is these variations which determine their characteristics such as hue and general appearance; they may even decide their ultimate value in the particular building situation.

The three main classes are:
Igneous rocks
Sedimentary rocks
Metamorphic rocks

Igneous rocks

The igneous rocks have been formed from molten material as it cooled and solidified. Their texture depends on the rate at which the cooling progressed since this determines the size and form of the crystals. Those which resulted from gradual crystallisation are composed of larger, more perfect, crystals and have the coarser texture. When solidified beneath the surface of the earth from magma, as the molten material is called, they are termed intrusive rocks, while extrusive rocks are those formed on the surface from lava, or volcanic fragments. Consideration of the igneous rocks is simplified if they are divided into three groups:
Plutonic rocks
Hypabyssal rocks
Volcanic rocks

(a) *Plutonic rocks*
The plutonic rocks were formed from magma buried at a great depth below the surface of the earth and consequently crystallised

very slowly. They have, as expected, the coarsest texture. Outcrops occur wherever erosion has removed the top strata.

The most common examples are the granites which are found in Cornwall and Devon, Westmorland and around Kircudbright and Aberdeen (the 'Granite City').

Quartz, a form of silica, is the main constituent, together with varying amounts of feldspar, a form of aluminium silicate. Other constituents include biotite, which is a black form of mica, an iron or magnesium silicate, and hornblende, calcium, magnesium or iron silicate. The crystal size of these minerals is such that they can be distinguished by the naked eye and, since feldspar itself may be coloured, the mottled appearance and colour of some granites is readily accounted for. Cornish granite, for example, is recognised by the prominent crystals of white feldspar which it contains, Shap granite contains pink feldspar in large crystals set among smaller ones due to crystallisation at different depths (porphyritic granite), and an even textured pink granite is found in Eastern Scotland.

(b) *Hypabyssal rocks*

These are formed by the injection of the molten material into crevices in the surrounding rocks where it cools more rapidly resulting in rocks of finer texture than the plutonic rocks.

An example is dolerite which occurs in *sills*—sheets which have intruded under a thin layer of other rocks. The Great Whin Sill has been formed in this way. It runs from the Pennines above the Vale of Eden through Teesdale and then north-eastwards to Bamburgh and the Farne Islands. In Teesdale it is visible at the Cauldron Snout and High Force waterfalls.

It is valuable as a road-making aggregate but it is not a favoured building material because of its dark colour.

(c) *Volcanic rocks*

The volcanic rocks were formed from the molten material which was forced up to the earth's surface and extruded as lava. Rapid cooling produced a rock which was composed of very fine crystals or was even non-crystalline.

The most common example is basalt, a green or dark blue, heavy, fine-grained rock consisting largely of feldspar and containing iron and magnesium. The grains can only be distinguished under a microscope.

Its main use is in road-making, since it is too dark and difficult to work for building purposes.

Sedimentary rocks

The sedimentary rocks have their origin in the erosion and disintegration of older rocks, the collection and layering of organic remains, and the precipitation of salts from solution.

Those which are composed of fragments of older rocks are called 'Clastic' or 'Detrital' rocks. Sandstone and shale are of this type and are, respectively, grains of sand and mud which have been cemented together.

Water is an extremely powerful solvent and as the streams and rivers make their way to the sea they are constantly searching out and dissolving any salts which lie in their path, and the sea itself performs the same action as erosion exposes the soluble material. All salts are soluble to some extent even though the solubility in some cases is only slight; this means that the sea is, potentially, a source of immense material wealth. Unfortunately no economically feasible means have yet been discovered whereby mankind can take advantage of the fact. The scale of these processes is brought out in an estimate that the erosion of land by streams lowers the entire land surface by one metre in 90000 years. Precipitation occurs when the solution becomes saturated as it would, for example, when a stretch of sea is isolated from the main body and evaporates.

Both types of sediment are converted into rock by *lithification* which involves compaction and then cementation in which the water is squeezed from between the particles and replaced by some material capable of binding the particles together. The most common binding materials are calcium carbonate, silica, and iron oxide.

Sedimentary rocks are the most widespread, underlying most of the earth's surface; metamorphic rocks are second in order of occurrence, whilst the igneous rocks are restricted to peculiar geological environments. As would be expected from the way in which the sediments accumulate in layers the sedimentary rocks have marked stratification with frequent changes in colour, texture, and mineral content.

Many of the sedimentary rocks are not suitable for use as building stones. The nature of the grains, the type of cement holding them together, and the extent to which the cementing action has progressed all play their part in deciding the properties of the rock. The sandstones and limestones are the two most important sedimentary rocks for building purposes.

(a)   *Sandstone*

Sandstone is composed essentially of quartz with minor amounts of other igneous fragments such as feldspar, mica, and hornblende. There are a number of different sandstones named according to the kind of material cementing the particles together. Since the quartz has excellent weathering properties the durability of the stone is largely determined by the matrix constituents and so such a nomenclature gives a guide to the weathering characteristics of the stone (Table 23).

Name of sandstone	*Cementitious material*
Argillaceous sandstone	Clay
Calcareous sandstone	Calcium carbonate
Ferruginous sandstone	Iron oxide
Siliceous sandstone	Silica

Table 23   A selection of types of sandstone

Siliceous standstones are the strongest and most durable. Two other names which are often given to sandstones are 'Arkose' and 'Greywacke'. The former are found on land and carry a considerable amount of feldspar while the latter are off-shore sandstones carrying variable amounts of dark matter such as clay. There are a great number of different varieties and the variations include not only stratified types but also some which are not stratified to any major extent: these are called 'freestones', possessing the useful attribute that they can be cut in any direction.

(b)   *Limestone*

Limestone has been formed from calcium carbonate originally in solution. The precipitation may have been caused in two ways, chemically or by the action of plants and animals. For example the carbon dioxide content of water is reduced by a temperature rise; it is also extracted by aquatic green plants. This causes a reduction in the amount of calcium bicarbonate which can remain in solution and so the calcium is precipitated as calcium carbonate. Deposits also arise from shells and remains or organisms such as coral which, during their lifespan, take up calcium carbonate to strengthen their skeletons.

The individual particles were compacted and cemented together in the same way as the quartz grains in sandstone. In this case, as would be expected, the main matrix-forming material is calcium carbonate.

Limestones contain varying amounts of magnesium carbonate and when the proportion is high they are called 'magnesian limestones'. Dolomite contains magnesium and calcium carbonates combined together as the double salt.

Limestone is an extremely valuable building stone and of the various types the most important are the oolitic limestones. These are made up of rounded grains reminiscent of fish roe. (The word 'oolitic' is derived from two Greek words, ōion meaning an egg, and lithos meaning a stone.) Portland stone is of this type and finds widespread application where resistance to the chemical weathering suffered in cities and industrial areas is demanded. It is a *freestone*. During the quarrying operations many beautiful and interesting fossils have been found, some of which are exhibited in the fascinating little museum at Portland—a place well worthy of a visit. Other oolitic limestones, Bath stone and Tayntonstone are examples, are also widely used but do not equal the weathering properties of Portland stone. The colour varies from white, in some Portland stone, through cream, in Bath stone, to shades of brown depending on the iron content.

Being a sedimentary rock, limestone is laid down in strata and care must be taken in selection since the quality of the stone is not necessarily the same in each bed.

Chalk is chemically identical to limestone but being much softer and porous is not used as a building stone.

Metamorphic rocks

The metamorphic rocks were formed from igneous or sedimentary rocks as they have undergone changes arising from movement of the earth's crust, heat, pressure, or chemical action. The extent of the changes depends on the type of rock and the particular circumstance to which it has been exposed. Generally several of these factors were at work simultaneously producing a rock which is significantly different from its parent. As an example, consider the effect of different environments on feldspar, a group of igneous rocks mentioned previously consisting of crystalline silicates of aluminium together with those of sodium, potassium, or barium. When a feldspar is exposed at the earth's surface it readily weathers to clay, a fine textured sedimentary deposit. If this clay becomes buried under layers of sediment it may be subjected over a lengthy period to a high pressure and temperature resulting in the formation of a number of contrasting metamorphic materials such as garnet, a semi-precious stone some species of which are most attractive gems rivalling rubies, or Muscovite, a white mica which is used as an insulator and as a lubricant.

Muscovite exhibits a feature characteristic of metamorphic rocks in having a *foliated* structure—being arranged as leaves or flat plates—in which the foliation planes grow at right angles to the direction of the pressure. In some metamorphic rocks these planes correspond to the original bedding planes but this is not always the case.

The two most important building stones in this class are slate and marble.

(a) *Slate*

Slate is formed by the action of intense heat and pressure on mud or clay. The heat first produces shale which is too weak to be used as a building stone but has been used as a retarder for Portland cement. The pressure changes the shale into slate by orientating the flaky grains into perpendicularity with the direction of the pressure and in doing so confers lines of cleavage enabling many slates to be split into thin wide slabs.

Slate is dense, durable, fire resistant, strong, and a good electrical insulator. It is widespread throughout mountain regions: Wales, the Lake District, Scotland, Cornwall, and Devon. It is quarried, wire sawn, and then split. Grey is the colour generally connected with slate because of the extensive use of this colour in roofing but it is available in tones of red, green, purple, and brown, which probably accounts for its increasing popularity for interior wall and decorative features.

(b) *Marble*

Pure marble consists of limestone which has been subjected to intense heat and pressure and allowed to recrystallise very slowly.

Most of the marble used in Great Britain is imported from France, Belgium, Sweden, North Africa, Greece, and Italy. The most prized marbles come from Greece and Italy and are available in a wonderful variety of colours and patterns. The colours are due to impurities in the original sedimentary rock. During the formation of marble the heat drives some of the carbon dioxide from the limestone and the remaining calcium oxide combines with any silica and other impurities such as iron to give the coloured veined and mottled effects.

Marble does not have the parallel structure possessed by many of the metamorphic rocks but has a *compact* or *massive* structure; the crystalline grains are so small that they cannot be distinguished except under a microscope. Slate and marble therefore require different working techniques.

Weathering of building stone

The weathering of natural building stone is dealt with in the Building Research Station Special Report No. 18. This is a most interesting report and is recommended for further reading.

Decay and disintegration of building stone is due mainly to chemical attack by gases present in the atmosphere and by temperature fluctuations, especially in the presence of water.

Atmospheric carbon dioxide (CO_2) dissolves in water to form carbonic acid which attacks the calcium carbonate present in limestone and calcareous sandstone—

$$CaCO_3 + H_2CO_3 = Ca(HCO_3)_2$$

The calcium bicarbonate is then washed away and, in the case of sandstone, the individual grains of sand are no longer cemented together.

Atmospheric sulphur dioxide (SO_2) likewise dissolves in water and attacks calcium carbonate. Here calcium sulphite is formed—

$$CaCO_3 + H_2SO_3 = CaSO_3 + H_2CO_3$$

The calcium sulphite then combines with atmospheric oxygen to form calcium sulphate which crystallises from solution as gypsum ($CaSO_4 2H_2O$). In addition the carbonic acid released continues the attack described in the last paragraph.

When sulphur trioxide (SO_3) is present in the atmosphere it dissolves in water to form sulphuric acid (H_2SO_4) which produces calcium sulphate directly from calcium carbonate.

The sulphur gases are only present to any serious extent in large towns and industrial areas. The Clean Air Act should do much to reduce the damage to stonework in these localities.

The stone, although incorporated into a building, is still liable to attack by some of the physical forces which caused the erosion in its natural state. Temperature variations, for example, cause the stone to expand and contract; the action of frost also weakens the stone. The movement of water by capillary action allows salts in solution to be transferred within the structure. The salts may then crystallise out as the water evaporates. Sandstone is particularly prone to this sort of damage.

Factors which influence the rate of weathering are:
Natural defects in the material.
Faulty craftsmanship.
Errors in the choice of materials.

(a) *Natural defects in the material*

(i) *Soft beds*
It was mentioned earlier that sedimentary rocks form in layers which are unlikely to have the same colour, texture, and mineral content. The weathering characteristics will therefore vary, depending on the layer from which the stone has been taken; and a quarry containing several different layers may yield several qualities of stone. Frequently the various layers are distinguishable and the *soft beds*—those with poor weathering properties—can be avoided.

(ii) *Vents and shakes*
A more serious defect is the presence of fissures in the stone arising, generally, from geological movements. The fissures may be so small that they are not noticed until the stone is worked or weathered. Sometimes calcite has been deposited in the fissures, having a cementing effect. The term *vent* is applied to a fissure which is prone to weathering and their presence may have serious consequences especially in decorative sections.

The term *shake* is applied to a fissure which has been so strongly sealed with calcite that the weathering property of the stone has been maintained. In limestone buildings shakes can often be seen quite plainly where they have weathered more slowly than the host limestone.

(b)Faulty craftsmanship

(i)Bedding

It is recommended that stones which have a laminated structure, for example sedimentary rocks, should be laid with the laminae horizontal. This inhibits any tendency for the laminae to separate when exposed to the weather and is called 'natural bedding'.

In special situations, such as cornices, parapets, and string courses, placing the stone with the laminae vertical and parallel to the vertical joints is preferred to avoid the loss of mouldings and throatings. This 'edge bedding' (alternatively termed 'joint bedding') introduces no loss of weathering resistance provided it is not used at corners.

If the stone is laid with the laminae vertical and parallel to the vertical joints the weathering properties are greatly impaired. This method of bedding, known as 'face bedding', must never be employed. It is sometimes the result of lack of care in determining the natural bed of the stone but it may be that the stone is not big enough for the course depth unless it is laid this way. Another temptation is that the stone dresses more smoothly parallel to the laminae.

Care must be taken to ensure that the stone is bedded evenly. If it is cut inaccurately, carelessly set, or there are pebbles in the mortar, the stone will be unevenly stressed and liable to spall.

(ii)Seasoning

Freshly quarried stone contains a certain amount of moisture, known as 'quarry sap', which should be allowed to dry out before the stone is used. There are a number of reasons for this:

To avoid frost damage.

If unseasoned stone is used the water it contains may freeze in the winter and cause disruption of the stone.

To avoid flaking.

It has been discovered that if the stone is carved while it is still *green* the surface may flake away in use.

To avoid decay.

During seasoning most of the salts present in quarry sap concentrate at the surface and can be removed in the working. This reduces decay which is attributed to the crystallisation of salts within the stone.

(iii)   *Quarrying and dressing*

The selection of good quality stone is a measure of the experience of the quarry-men but even when this has been done it may be marred if the quarrying and subsequent dressing are not executed with care.

The major points in this regard are:

Blasting.

When blasting it is important that only moderate charges of powder be used to avoid cracking the stone. Even minute cracks will take in water and reduce the useful life of the stone.

Bruising.

Hammer dressing, careless machine dressing, the use of blunt tools, or rough treatment of the stone after working, give the stone a *bruise* which is likely to spall off in time.

(c)   *Errors in the choice of materials*

(i)   *Iron dowels*

Dowels and cramps should be made from non-corrosive metals such as alloy steels or alloys of copper or nickel.

If iron fittings are used the expansion arising from corrosion causes vents to appear in the stone and increases the rate of weathering.

(ii)   *Pointing*

Excessive deterioration of stonework is caused by the application of dense Portland cement mortars.

(iii)   *Incompatible materials*

Care must be taken to avoid the possibility of interaction of different types of stone. Sandstone and limestone must be kept separate. As limestone weathers calcium sulphate is produced which may be washed into the pores of the sandstone where it crystallises out of solution causing disintegration. Likewise limestone and magnesian limestone should not be used together since magnesium sulphate causes accelerated disintegration of the limestone.

Questions Please see appendix at end of book.

Further reading

Building Research Station, Special Report No. 18 (H.M.S.O.).

Seven   Introduction to Plastics in Building

More than 24% of the tonnage of plastics consumption in the U.K. is now accounted for by the building industry. In 1962 building industry applications consumed 90000 tonnes, in 1965 the figure had risen to 170000 tonnes for the year, and by 1970 it is estimated that the 300000 tonnes mark has been exceeded (*see* Fig 29).

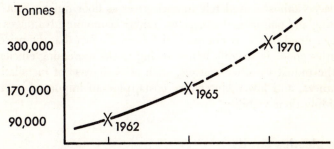

Fig 29   Growth of plastics consumption in the building industry

This is a growth rate of about 15% per annum and is quite impressive when compared with about 3% per annum for the building industry as a whole over the same period. However, compared with the potential which exists for the use of plastics materials in building, the consumption figures are low. A suggested reason for this failure to realise the potential is given by an architect writing in the sales brochure of one of our leading plastics manufacturers.

'Nothing has been gained from that era when the sales promoters clambered over each other to befriend the interested architects and major building contractors; nor from the misplaced enthusiasm of young architects and fabricators who impetuously published their irrational theses on plug-in cities and raved their curvilinear panaceas for the nation's ills while

they ignored the pleas of their respective professional and commercial parent bodies to assist in the mundane tasks of getting
the inglorious products and data evaluation exercises off the
ground. Similarly, nothing has been gained from expositions
that the physical and financial structure of the building industry
is all wrong and that an adoption of an economics framework,
similar to that practised in other commercial industries, would
remould the industry to respond, automatically, to the all-
plastics innovators.'

There was a time when plastics were considered to be a 'cheap
and nasty' substitute for the 'real thing'. The technological compe-
tence and integrity of the modern plastics industry have gradually
changed this attitude and proved that, in many areas, plastics are
technically superior to traditional materials; the economic advantages can be regarded as a bonus. Traditional materials have been
replaced almost completely in such areas as floor finishes, surface
coating and paints, and adhesives. Where economic advantages are
to be gained they may not only arise from the fact that the price of
plastics goods is steadily falling owing to the increasing efficiency
of the industry; other reasons, such as their ease of installation,
lightness, and low maintenance costs, play an important part in
deciding their viability.

The plastics family

The term 'plastics' covers those organic (i.e. carbon-based)
materials which at some stage during their manufacture possessed
the property of *plasticity*, enabling them to be shaped, usually by
the action of heat and pressure.

Some plastics can be softened by heating, whereas others are
unaffected unless they are heated to a sufficiently high temperature
to cause decomposition. This provides a basis for a broad classification into *thermoplastics* and *thermosetting plastics*.

Thermoplastics
A thermoplastic material softens and sometimes melts when
heated but hardens again when allowed to regain its former
temperature.

Examples:
   Polythene

Polypropylene
Polyvinylchlride (P.V.C.)
Polystyrene
Nylon
Acrylics, e.g. Perspex

Thermosetting plastics
A thermosetting plastic does not soften when heated and retains its shape.
Examples:
Epoxy resins
Polyester resins
Synthetic rubbers
Polyurethanes
Phenolics, e.g. Bakelite
Urea and melamine formaldehyde resins
The thermosetting plastics are generally more solvent-resistant than the thermoplastics but are not so flexible.

Polymerisation

Plastics are manufactured by causing small organic molecules—monomers—to link together to form very large molecules—polymers—by a process known as polymerisation.

Polymers occur naturally; casein, wool, hair, silk, and natural rubber are examples; but the majority of those used in building are *synthetic*.

The structure and composition of these large molecules are widely variable, offering a vast range of properties:
(a)   The number of monomers linked together can be controlled; polythene, for example, may have a chain length of 800 units.
(b)   They may be built up either in long chains or in two- or three-dimensional networks. Thermoplastics are of the long chain type. Thermosetting resins have reactive groups along the chains and *cross-link* during the heat treatment to form a rigid structure, making them more brittle than the thermoplastics.
(c)   The monomers may be identical or mixed; for example the A.B.S. resin is synthesised from three monomers, acrylonitrile, butadiene, and styrene.

Polymerisation can proceed by two mechanisms.

Addition polymerisation

This is a process involving *unsaturated* molecules. We have said that the plastics are derived from organic compounds. These all contain carbon, which has a valency of four. This means that it can form four chemical bonds. If all the valencies are satisfied as, for example, in methane (CH_4) the product is *saturated*. If only three are satisfied the product contains a double carbon to carbon bond as in ethylene ($CH_2 = CH_2$). If only two are satisfied then a triple bond is required, as in acetylene ($CH = CH$). The compounds containing multiple bonds are relatively unstable and under suitable conditions these bonds open up producing a free valency which joins on to similarly produced free valencies.

Example:

P.V.C.

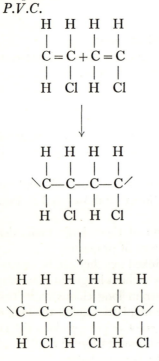

Two vinyl chloride monomers

Under the influence of a catalyst to make the reaction quicker

The double bonds open and the monomers join
There are still 2 valencies unsatisfied

The double bonds of another monomer open and it joins on
There are still 2 valencies unsatisfied and so the reaction proceeds.

This chain reaction continues until it is terminated by the introduction of an impurity to satisfy the terminal valencies. The full reaction is written—

$$2n(CH_2 = CHCl) \rightarrow (\cdots -CH_2-CH-CH_2-CH-\cdots)n$$

$$\underset{Cl}{|} \qquad \underset{Cl}{|}$$

2n vinyl chloride   polyvinyl chloride
    monomers   polymer

where n is the number of repeating units in the chain.

Condensation polymerisation
Condensation polymerisation consists of the joining together of
two molecules with the elimination of a simple molecule such as
water or ammonia. The product must still be capable of reacting
with more monomers to continue the polymerisation.

Example:
Nylon.

Water being eliminated

H   H HO   OH
 \N—C—(C$_4$H$_8$)—C—N⁄ + ⁄C—C—(C$_2$H$_4$)—C—C⁄
H⁄   \H   O   \O

Hexamethylene diamine   Adipic acid
        (H.M.D.)

H   OH
 \N—C—(C$_4$H$_8$)—C—N—C—C—(C$_2$H$_4$)—C—C⁄   + H$_2$O
H⁄ | || \O
             H O   Water

The H.M.D. contains terminal amino groups $\left(-N{<}^H_H\right)$ which

combine with the terminal carboxyl groups $\left(-C{<}^{OH}_O\right)$ of the
adipic acid to eliminate water.

This proceeds in the same way to produce the polymer nylon
the characteristics of which depend on the number of units in the
chain and how they are linked.

Sources of raw materials

By far the greatest supplier of the raw materials for the production
of plastics is the petrochemical industry. Secondary sources are
coal tar and natural gas. This latter source is expected to grow in

importance as the North Sea fulfills its promise. Table 24 lists some of the sources.

Material	Source	Products
Acetylene	Cracking* of natural gas	Neoprene, P.V.C., polyvinyl acetate (P.V.A.)
Ethylene	Refinery gas or cracking of crude oil products	Polythene, polystyrene
Propylene	Cracking of crude oil products	Polypropylene, acrylics, phenolics
Hydrocarbons, e.g. Butane	Cracking of crude oil products	Nylon, rubber, A.B.S. resin
Benzene	Distillation of crude oil, coal	Polystyrene
Xylene	Distillation of crude oil, coal	Terylene
Urea	Ammonia, carbon dioxide	Urea formaldehyde resins
Formaldehyde	Air and coal	Urea and melamine formaldehyde resins. Phenol formaldehyde resins
Phenol	Distillation of coal tar. Synthesis from crude oil products	Phenol formaldehyde resins

* Cracking—The breaking down of large molecules into smaller ones by heating in the presence of steam or catalysts.

Table 24   Sources of raw materials for plastics manufacture

Processing of plastics

Plastics are manufactured in the form of liquids, powders, or granules, and require to be processed before they are useful to the builder. The particular means of processing must be appropriate to the nature of the resin, the form in which it is supplied, and the article to be fabricated from the processed material. There are several techniques from which the choice is made:

Extrusion
Moulding
Casting

Vacuum forming
Calendering
Laminating
Foaming

Extrusion

Granules are fed by a hopper into a heated barrel containing a revolving screw (Fig 30). The screw is so designed that it not only carries the plastic along the barrel but also squeezes out any air trapped in the melt.

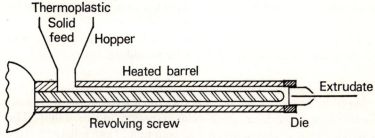

Fig 30   Extrusion

When the melt reaches the end of the barrel it is forced through a die, cooled, and cut to length. The shape of the die determines the profile of the extrudate.

Extrusion is virtually a continuous process, and an extremely versatile one being capable of producing rods, pipes, shaped profiles, fibres, sheets and film, and covered wire. When pipe or tube is being extruded it is necessary to prevent it collapsing. This is achieved by filling it with compressed air through a fitting in the die head. Wire is coated using a *cross-head* extruder in which the wire and the plastic pass through the die together. The extrusion process is also involved in some of the other methods described below.

Moulding

(a)   *Compression moulding*

The granules or powder are placed in the bottom half, the *die*, of a heated mould and the top half, the *plug*, is pressed into it. The whole is then held between the platens of a hydraulic press until sufficient time has elapsed for the plastic to have melted and flowed to fill every part of the mould. When thermo-setting resins are being moulded the time and temperature must be chosen to ensure that the resin is completely cured. After cooling the article is removed and polished to remove the excess, or *flash*, which has been forced out between the two halves of the mould.

(b) *Injection moulding*
Molten resin is injected into a cool mould using either an extruder or a piston operating in a heated cylinder.

Injection moulding is most suitable for thermoplastics although a variation, transfer moulding, where the mould is heated can be used to form thermosetting resins.

Light fittings, switches, door furniture, pipe fittings, and cisterns are all conveniently made by moulding.

(c) *Blow moulding*
Shaped hollow articles such as bottles or closed cisterns are manufactured by a combination of tube extrusion and moulding. The extruded tube is passed into a cold split mould and inflated to take the shape of the mould, using compressed air.

(d) *Slush moulding*
Articles such as P.V.C. cistern balls are readily made by slush moulding. A paste is run into the heated mould which is rotated until the P.V.C. gels and it is then cooled.

Casting
The liquid resin is poured into a mould and allowed to harden with the application of heat if necessary. Electrical components are *potted* by this method using liquid epoxy resins.

Vacuum forming
Thermoplastic sheet is shaped by heating it and drawing it into the mould using an air pump to produce a vacuum between the sheet and the mould (Fig 31).

Laminating
Laminating is the process of combining several layers into one composite sheet. Paper, fabric, or wood veneers are first impregnated with resin; the separate layers are then superimposed in a

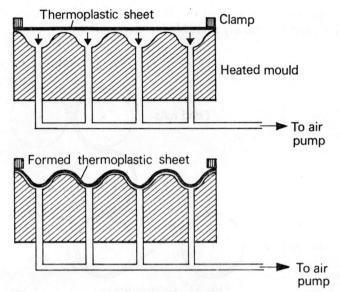

Fig 31   Vacuum forming

mould and the resin cured. The details of the curing procedure are dependent on the type of resin employed.

(a)   *Low pressure laminating*

Laminates based on polyesters or epoxy resins are obtained by holding the resin impregnated layers firmly in contact until curing is complete. Cold curing, accelerated by catalysts, is generally sufficient but in some cases moderate heating is employed.

Glass-reinforced plastics are made in this way.

(b)   *High-pressure laminating*

Laminates based on phenol formaldehyde, urea formaldehyde, or melamine formaldehyde resins are cured by heating. Since water is formed during the reaction it is necessary to apply higher pressure (approx. 8 MN/m²) to prevent *delamination*.

Decorative laminates, such as Formica and Warerite, are made in this way.

Calendering

Calendering is the process most favoured for the production of P.V.C. sheet especially when decorative effects, such as embossing,

F

are required. It is also an excellent way of coating fabrics with P.V.C. and is generally available for thermoplastics.

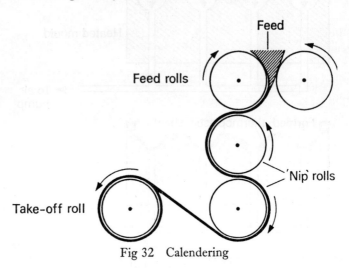

Fig 32   Calendering

A stiff dough from a pre-mixer is squeezed between heated rolls to produce a melt which is then passed between rolls whose distance apart determines the sheet thickness (Fig 32). It is then cooled over a final take-off roll.

Foaming
Foamed, or expanded, plastics are made by mechanical agitation to include air or by the use of chemical *blowing* agents.

Polystyrene can be obtained in the form of expandable beads containing a volatile liquid. They are processed by subjecting them to steam treatment to give pre-expanded beads which are then moulded in the presence of steam.

Polyurethane foams can be prepared by the addition of water to the basic ingredients of the polymerisation. It reacts with one of them to produce carbon dioxide gas. The bubbles of gas are imprisoned as the polymer grows and cross-links.

When the foams are intended for thermal insulation applications care is taken to ensure that the cells are *discrete*, each cell being separate from its neighbour. This closed cell structure prevents heat being conveyed by convection between one cell and the next.

Plastics having applications in building

The building industry is a tremendous growth area for plastics applications. Whilst not all plastics are suitable the range is steadily increasing as improvements and new developments appear. In this section we examine the plastics most widely used. The naming of plastics is dealt with in B.S. 3502: 1962 'List of Common Names and Abbreviations for Plastics'.

Acrylics, e.g. Perspex

Polymethyl methacrylate, the most common acrylic plastic material, is manufactured from the monomer methyl methacrylate. It is a thermoplastic and is available both in sheet form and as beads for injection moulding.

Being hard and transparent to light, and possessing good weathering properties, it is valuable for glazing, roof lighting, light fittings, and outdoor signs. It can be obtained in a wide range of colours and, since it is resistant to attack by alkalis, salts, and detergents, it can be used for bathroom suites and kitchen sink and drainer units.

It is also used in paint formulations and solvent based adhesives.

Acrylonitrile butadiene styrene (A.B.S.)

A.B.S. is a copolymer of acrylonitrile, butadiene, and styrene. These three monomers can also be reacted to form synthetic rubbers (nitrile rubber is acrylonitrile-butadiene) and so it is not surprising that A.B.S. resins are endowed with certain rubber-like properties—in particular, toughness and high impact strength. A broad range of properties can be obtained by varying the ratio of the constituent monomers. A.B.S. is a thermoplastic.

It is used to manufacture extruded and injection moulded articles such as waste pipes (good chemical resistance), pipe fittings, trays, and door furniture.

Amino plastics

The amino plastics in common use are urea and melamine formaldehyde. They are so-called since they contain an amino (NH_2) group in their chemical formula.

Urea formaldehyde (U.F.) resins have good electrical properties and moulded applications include plugs and sockets, but lamp

F*

holders are preferably made from melamine formaldehyde (M.F.) resins which are more heat stable. They are also used for door furniture and W.C. seats.

M.F. resins find their greatest application in laminates for door panels, kitchen work tops, and decorative purposes. There are also exterior grade laminate veneers popular in the shop-fitting trade. The laminates can also be bonded to cellular plastics and particle board.

The amino plastics are also used in adhesive formulations.

Epoxy resins, e.g. Araldite
The epoxy resins are mainly used in adhesives, reinforced plastics, surface coating, and concrete repair work. They cure when mixed with a 'hardener'. Flooring compositions suffer only slight shrinkage, are sufficiently hard for traffic within 24 hours, are highly resistant to attack by chemicals, and are tough.

Phenolic resins, e.g. Bakelite
The first phenolic resin to be made, and, incidentally, the first completely synthetic resin, was phenol formaldehyde (P.F.). It was called 'Bakelite' after L. Baekeland who discovered how to prepare it in suitable form by reacting phenol and formaldehyde in the presence of a catalyst. Since then other members of the phenol family, including cresol and resorcinol, have been used to prepare resins; they are all classed as phenolic resins, and are thermosetting.

Although phenol formaldehyde resins have high heat resistance they are discoloured by sunlight and consequently are only available in the darker colours. They are brittle and so fillers are added to improve their impact resistance. The most effective filler in this respect is chopped cotton fabric. Other fillers used include asbestos, wood flour, and minerals. P.F. resins are cheaper than U.F. and M.F. resins and are often used as the *core stock* of decorative laminates; only the top layer is composed of the more expensive resin.

The moulding applications include electrical fittings and door furniture.

P.F. resins are also used in adhesives and surface coatings.

Polyamides, e.g. nylon
The best-known polyamides are the nylons. Nylon 6,6, which is

prepared from adipic acid and hexamethylene diamine, can be obtained in a form suitable for spinning as well as in chip or granular form for moulding.

It is chemical-resistant, abrasion-resistant, tough, self-lubricating, light, and has a high softening point for a thermoplastic (over 200°C). Nylon can be pigmented and reinforced with glass fibre.

The main building applications are rollers for sliding doors, door catches, and cistern valve assemblies.

Polyesters

The polyesters are manufactured by heating a polyhydric alcohol (one containing two or more —OH groups) with a polybasic acid (one containing two or more —COOH groups).

$$n(HOOC—R—COOH) + n(HO—R'—OH)$$
$$\rightarrow (—OOC—R—COO—R')n + 2nH_2O$$

where R is the main part of the acid molecule.

R' is the main part of the alcohol molecule.

This process is known as 'polyesterification'.

It is possible to produce both thermoplastics and thermosets by this method depending on the nature of the monomers. If one of them is unsaturated, cross-linking will occur as the double bonds open to produce the thermoset resin. It will be appreciated, therefore, that the polyesters are a large family having a wide range of properties and applications.

Glass reinforced polyesters may be produced by lamination using glass fibre matt or by the incorporation of chopped fibre in the mix. They have, in general, high impact strength, high strength/weight ratio, good weathering properties, good chemical resistance, and surface hardness. Improvements are being made in their fire retardant properties to meet the stringent building regulations in this regard. The sheet is preferred to rigid P.V.C. for roof-lighting on industrial premises on account of its greater toughness and stiffness. Glass-reinforced polyesters are also used for dome lights, lighting panels, external ventilator mouldings, prefabricated building sections, and stressed skin roof structures.

Polyesters also find application in cement mixes for quick setting repair to concrete surfaces, in floor finishes, adhesives, and furniture finishes.

The alkyds are a well-known class of polyesters used as the binder in paint formulations.

Polythene, e.g. Alkathene

Polythene (polyethylene) can be obtained in a density range from 920 kg/m³ to 960 kg/m³ by blending a low-density and a high-density product or by adding other monomers to the polymerisation mixture. The two basic polythenes are produced, in the U.K., by a high-pressure and a low-pressure process respectively.

(a) *Low-density polythene—I.C.I. low-pressure process*

Low-density (L.D.) polythene was first produced by I.C.I. in 1935 by a *mass* polymerisation method. Pure ethylene is heated to 190°C under a pressure of 1500 to 2500 atmospheres in the presence of a small amount of oxygen which is a catalyst for the polymerisation. After removal of unreacted ethylene the polythene is extruded, chilled, and chopped into granules.

(b) *High-density polythene—Ziegler low-pressure process*

High-density (H.D.) polythene can be obtained by the Ziegler (1955) process which uses a complex metal-organic catalyst. The catalyst contacts the ethylene in the presence of an unreactive hydrocarbon solvent under 6 to 7 atmospheres and a temperature of 100 to 170°C. After removal of the solvent and catalyst the polythene is extruded, chilled, and chopped into granules as in the low-pressure process.

There is a steady gradation in properties as the density is increased: those of intermediate density material fall between the two extremes. L.D. polythene is tougher, softer, and less rigid than H.D. which has lower impact strength but is not softened by boiling water and is more abrasion-resistant. Both forms are excellent electrical insulators and both have low water permeability and water absorption. Their chemical resistance is generally good although they are attacked by particularly vigorous agents such as nitric acid.

Polythene is degraded by the ultra-violet component of sunlight and stabilisers are added during its manufacture; carbon black is used when colour is not important. Anti-oxidants are added to stabilise polythene from attack by atmospheric oxygen. Another stabilising agent is incorporated to combat a peculiar defect of polythene known as *environmental stress cracking*. This refers to the cracking which occurs when polythene is stressed in contact with certain chemicals, notably detergents.

Polythene burns readily when a flame is applied and continues to burn when the flame is removed; that is to say it is not 'self-extinguishing'.

Applications include cold water pipes and fittings and cold water cisterns. On building sites polythene is used for damp courses and for temporary weather protection. In concrete work when used as an underlay it prevents moisture draining out of the mix during curing and as an overlay it prevents cracking and spalling by ensuring a slow and steady cure. When reinforced concrete is laid on a concrete base a film of polythene placed between the two allows them to move relative to one another reducing the number of expansion joints required.

Polypropylene, e.g. Propathene

Polypropylene is manufactured by polymerising propylene using Ziegler type catalysts. It is a thermoplastic and is available both as sheet and as extruded granules for moulding.

It is lighter, harder, and glossier than polythene, and has a higher softening point (about 150°C) and abrasion resistance. Its electrical properties are very similar to those of polythene. A peculiar property possessed by polypropylene is referred to as the *hinge effect*. When a piece is bent over on itself a hinge is formed which will withstand continuous flexing for many months without snapping off. Anti-oxidants and U.V. stabilisers are incorporated during manufacture.

Polypropylene applications are similar to those for polythene but since it is more expensive it only replaces polythene where the article is required to withstand higher temperatures.

Polystyrene

Polystyrene is manufactured by polymerisation of styrene monomer which is produced by reacting ethylene with benzene. It is a thermoplastic and is available in granular form and also in the form of expandable beads. The basic polymer possesses low impact strength and other polymers are often incorporated to give a less brittle material. An example of this *toughened* polystyrene is the A.B.S. resin described earlier.

The main building applications concern expanded polystyrene which has a closed cell structure and is an excellent thermal insulating material. It is very water-resistant and so acts as a *vapour barrier*. Although resistant to many chemicals it is attacked by concentrated nitric acid and certain organic solvents such as acetone and esters used in nitrocellulose finishes. Expanded polystyrene is *rigid* and so will not recover if crushed. Normal grades support combustion presenting a fire risk. It is also of value in acoustical correction of buildings being a good sound absorber of the porous type. It is used in the form of wall and ceiling tiles, wall-paper backing, and *sandwich* type laminates.

The translucency of polystyrene sheet enables it to be used in the construction of diffusers for fluorescent lighting installations and as panels in suspended ceilings.

Polyurethanes

Polyurethanes are prepared by reacting a di-isocyanate with a compound containing two or more hydroxyl (—OH) groups. A cross-linking agent is then added.

Lacquers can be prepared by dissolving them in an organic solvent or they can be foamed, in the manner described earlier.

The foams can either be rigid or flexible depending on the nature of the ingredients and can have either an open or closed cell structure. They can be made self-extinguishing although they will burn whilst in contact with a flame.

The flexible foams with an open cell structure are valuable acoustic materials absorbing most strongly in the high frequency range. They are also used in draught excluder strip and in moulded cushioning.

Rigid foams with a closed cell structure are excellent thermal insulating materials. They can be foamed *in situ* within cavity walls. In an existing building 20 mm diameter holes are drilled in the mortar courses at 1·2 m intervals and the liquid ingredients injected. The foam sets almost immediately. It is claimed that 'Ufoam' (I.C.I.), once in place, is unaffected by frost, temperatures up to 130°C, and vibration; it resists water penetration, mould, rot, vermin, and pests. In an average three-bedroomed house such treatment makes it possible to reduce the design rated output of the heating system by 20 to 30%.

The foams have good adhesion to wood, metal, and brick and have been made the core of sandwich panels between, for example, aluminium faced plasterboard. They have a high strength/weight ratio.

Polyurethane surface coatings are hard and durable, adhere well, and have a high gloss retention.

Vinyl plastics

The vinyl plastics most used in the building industry are poly-vinylacetate (P.V.A.) and polyvinyl chloride (P.V.C.). The vinyl radical ($CH_2 = CH_2$—) is derived from acetylene ($CH = CH$) in which the unsaturation has been reduced. The monomers vinyl acetate and vinyl chloride are variations on a theme which has been mentioned before and is summarised in Table 25.

Monomer	Chemical structure	Interrelation	Polymer
Ethylene	$CH_2 = CH_2$	Vinyl radical attached to a hydrogen atom	Polythene
Vinyl acetate	$CH_2 = CH.OOCH_3$	Vinyl radical attached to an acetate radical	P.V.A.
Vinyl chloride	$CH_2 = CHCl$	Vinyl radical attached to a chlorine atom	P.V.C.
Styrene	$CH_2 = CH.C_6H_5$	Vinyl radical attached to a benzene ring	Polystyrene
Methyl methacrylate	$CH_2 = CH.CH_3COOCH_3$	Vinyl radical in which one hydrogen atom has been replaced	Perspex

Table 25   Polymers derived from vinyl unsaturation

P.V.A. is a thermoplastic but is not suitable for moulding. It is generally emulsified in water and used as a paint or an adhesive, after stabilisation and pigmenting. The emulsions are also mixed with Portland cement to provide resilient finishes for application to concrete floors. Although oil and grease-resistant they are slightly porous and require to be wax polished at intervals.

P.V.C. is a thermoplastic available in both rigid (unplasticised) and flexible (plasticised) forms.

(a) *Rigid P.V.C.*

Rigid P.V.C. is manufactured as a white powder and is processed into extrusions, mouldings, sheets, and cellular form. It is tough, weathers well, can be pigmented in a wide colour range, is self-extinguishing (i.e. does not continue to burn when the flame has been removed), and is an excellent electrical insulator.

The building applications include rainwater goods and pipes for cold-water plumbing and drainage. There are many advantages when plastics are used:

(a)   Lightness; one man can carry 37 m of P.V.C. guttering compared with 7·5 m of cast-iron guttering.

(b)   Fewer joints are required since longer, extruded, lengths are available.

(c)   No jointing compounds or special tools are required. Jointing is often accomplished by means of 'O' rings and sometimes the material is welded.

(d)   The lengths can be cut to size easily, using a hacksaw, and no matching up of sockets and spigots is necessary.

(e)   They are corrosion-resistant.

(f)   They are self-coloured. No painting is required.

(g)   The smoother bore of pipes means that the pressure drop is smaller and the flow is improved.

P.V.C. is an extremely versatile material and the list of applications is very long indeed. Being obtainable in transparent, translucent, and opaque grades it is useful, for example, in lighting installations ranging from roof-light sheet, and illuminated ceilings, to reflectors for use in kitchens and factories where corrosion would otherwise be a problem. Doors and exterior wall cladding are contrasting applications. Steel and aluminium have been coated with P.V.C. for use in roofing, partitioning, and the manufacture of window frames.

(b) *Flexible P.V.C.*

The P.V.C. powder is mixed with a plasticiser to form a paste which is then heat cured. Not only does the addition of the plasticiser endow flexibility but it also permits extrusion at a lower temperature, reducing the risk of the polymer breaking down chemically.

The properties of plasticised P.V.C. can be tailored to suit the end use by careful selection of the filler content and type. Consequently flexible P.V.C. is used widely in building.

P.V.C. floor tiles and continuous flooring have good resistance to abrasion, chemicals, fire, and heat. Antistatic additives improve the electrical conductivity to prevent the accumulation of dust. It is not possible to give a full list of applications here but they include waterproof membranes, hose-pipes, supported film for decorative purposes, and covered wire for both fencing and electrical purposes.

Paints based on P.V.C. can be applied direct to brickwork, concrete, plaster, plasterboard, and wood wool slab. They are durable, easily cleaned, warm to the touch, and sufficiently elastic to maintain a continuous film as the substrate expands and contracts.

Questions

1 Plastics are manufactured by a 'polymerisation' process. What does this term mean?

2 Describe briefly the mechanism of one polymerisation process giving in your answer the name of a plastic material manufactured by the process selected.

3 Plastics are manufactured in the form of liquids, powders, or granules. List the techniques available for processing them into a form useful to the builder.

4 Describe one of these methods and name a material which may be processed by the technique selected.

5 State the essential differences between a thermosetting and a thermoplastic plastic. Give one example of the use of each type in building, referring to the properties of the material which make it particularly suitable for the use to which it is being put.

6 Using the following list of plastics prepare a table showing their major uses in building and the properties which make them suitable for the particular applications:

Acrylics
Amino plastics
Epoxy resins
Phenolic resins
Polyamides
Polyesters
Polyolefines
Polyurethanes
Vinyl plastics

7 Below are listed several areas in building wherein plastics may

F**

be used. What are the likely advantages and disadvantages of their use compared with the use of traditional materials?

> Floors
> Walls
> Roofs and ceilings
> Plumbing
> Building sites

Your answer should include mention of the plastics and the traditional materials they replace.

Eight   Paints

Coatings of paint are applied to nearly every type of article and structure. These coatings provide protection against corrosion, increase the resistance to hard wear, and give the item or structure a more pleasing appearance. A paint is a composition which is applied in a liquid or plastic condition and subsequently hardens to form a solid coating. Paints vary widely in composition and, nowadays, the range of natural and synthetic (man-made) materials used in paint manufacture is so great that a simple classification becomes very difficult. For good results, it is therefore important that the user should take note of the manufacturer's instructions concerning the suitability of the paint, the preparation of the surfaces, and the method of application.

What is paint?
Essentially, a paint consists of a suspension of fine pigments in a liquid medium, sometimes called the vehicle. A varnish does not contain pigments.

Pigments
The solid particles in paint are called pigments. The pigment is chosen because it possesses certain properties or a particular combination of properties, e.g. anti-corrosive properties, as mentioned in Chapter Nine. It may be used because it confers some or all of the following properties on the paint film:
(a)   Colour.
(b)   Obliterating power, i.e. covers previous colours.
(c)   Greater strength and adhesion.
(d)   Improved durability and weathering properties.
(e)   Reduced gloss.
(f)   Modified flow and application properties.

Types of pigment
Pigments are either:
Synthetic, i.e. man-made,
or natural and organic or inorganic chemicals.

Organic pigments are those which have a plant or animal origin although, strictly speaking, they are all compounds of carbon. The remainder are called inorganic. Some of the newer pigments contain metallic elements and organic structures. Generally speaking, the organic pigments provide the most attractive cleanest colours. On the other hand, the inorganic pigments are more stable to light, resistant to heat, and give the best white and black paints. The purest white pigment is titanium oxide and the most jet black is carbon, which is considered to be inorganic. The best organic or organometallic pigments give the highest standard of performance for most uses, e.g. 'Monastral' pigments and 'Cinquasia' pigments.

Thus, the choice of pigment for a particular paint is decided by a consideration of the above factors.

It should be remembered that pigments are either naturally occurring compounds of organic or inorganic origin, or they are artificially made materials. Thus artificial white pigments have greater opacity than the natural whites but both might be used in the same paint because this makes application easier. Alternatively, the natural pigment may assist the particles to remain in suspension more readily, in which case it is called an *extender*. The production of paints of particular colour is determined by mixing suitable pigments.

The medium or vehicle
The desirable properties of a vehicle are:
(a) It thoroughly wets the particles of the pigment.
(b) It should be sufficiently viscous to hold the particles in suspension when the paint is drying, but sufficiently fluid to make application easy.
(c) It should form an adherent tough elastic film on the surface to which it is applied.

To achieve these properties, the vehicle consists of a number of different substances. Substances found in the vehicle include the following:

Thinner
e.g. turpentine or white spirit (a petroleum distillate). This gives

increased mobility to the paint. The thinner has to dissolve the main body of the paint and then evaporate, leaving it chemically unchanged.

Gum

A gum is a carbohydrate, i.e. a compound of carbon, hydrogen, and oxygen, which is soluble in water, e.g. gum arabic, which is a tree extract. A gum is an example of a binder, which gives increased adhesion to the painted surface.

Resin

A resin may be either naturally occurring—produced by certain trees or insects—or be synthetic, i.e. man-made. Resins are complex substances which harden on exposure to air by a process which involves oxidation, evaporation, and polymerisation. Polymerisation is the joining together of the molecules of a substance to form larger molecules. A resin gives additional gloss, bulk, and toughness to the paint film.

It appears that one of the main advantages of synthetic resins over natural resins is that the quality of the synthetic resin can be carefully controlled during its manufacture whereas the quality of the natural resin varies widely depending on its origin. Many types of synthetic resin are now used in paints; these include coumarone, polystyrene, epoxy, vinyl, urea, polyurethane, and alkyd resins.

Alkyd resins

These are all made from phthalic anhydride and glycerol, the product being modified with linseed, soya bean, or castor oil. They are classified according to their oil content. Thus, there are long oil alkyds, which are used in primers, undercoats, and gloss finishes; medium oil alkyds which are used in industrial finishes and stoving paints; and short oil alkyds which are used in spraying and dipping finishing paints. All the alkyd resins have good adhesion and help to produce elastic and lustrous finishes. The name 'alkyd' comes from its ingredients, namely alcohol and acid.

Coumarone resins

These are used in the manufacture of primers for plasters and cements because, when mixed with tung oil, they are very resistant to alkalis. The resins are hard, brittle, and yellow in colour and are made by polymerising certain constituents of naphtha, chiefly coumarone and indene. Naphtha is the mixture of hydrocarbons which may be obtained by the distillation of coal tar. Coumarone resins are also used in metallic bronze paints and cellulose finishes.

Epoxy resins

These are extremely versatile resins which produce coatings that are very resistant to chemical attack, give good adhesion, and produce tough flexible paint films. They do, however, give gloss finishes which have a tendency to chalk, and the resins are relatively expensive. An epoxy resin is a substance which contains a maximum of two epoxide rings at the ends of the molecule and a number of hydroxy groups in the molecule. They may be combined with other resins to give, for example, epoxy alkyd or epoxy phenolic finishes. (The epoxy group is $-CH-CH_2$, the hydroxyl group $-OH$.)

$$\begin{array}{c} \diagdown \quad \diagup \\ O \end{array}$$

Polyurethane resins

These produce paints which are hard, tough, abrasion-resistant, and give good resistance to atmospheric and marine attack. All polyurethane finishes contain compounds called isocyanates (which contain the grouping $-NCO$) and contain urethane linkages of formula

$$-N-C-O- \\ \ \ \ | \ \ \ \| \\ \ \ H \ \ O$$

They are made by polymerisation of organic di- or poly-isocyanates with dihydric alcohols.

The brief details given of four of the types of resin commonly used should help to illustrate the importance of the resins as a group of materials. It should also be realised that there are many other resins used in paints.

Drying oils

These are mainly the *fatty* oils, which are largely vegetable oils extracted from the seeds or the fruit of many types of vegetable matter. Drying oils form a film, i.e. harden, on exposure to air. Examples are linseed oil which is extracted from the seeds of the flax plant and tung oil which comes from the seeds of the Aleurites tree. Drying oils contain a proportion of fatty acids containing three double bonds combined with glycerol.

To reduce the hardening time, these oils are normally processed before use giving, for example, boiled linseed oil, stand oil, and blown linseed oil. Stand oil is oil heated to 230° to 260°C in the absence of oxygen. Blown oil is oil with air bubbled through it at 75° to 120°C.

The drying time of these oils may be further reduced by the addition of *driers*.

A drier is a metal soap with an acid part which assists the solvent powers of the oil medium. The metals involved are lead, cobalt, and manganese.

Manganese linoleate is an example of a metallic soap. In order to achieve uniform drying of the paint film, a mixture of metal soaps is used. The driers normally comprise less than 1% of the paint.

Plasticisers
A plasticiser is a solvent for the resin or polymer. A plasticiser is used to improve the flexibility of the hardened paint film. It can also improve the adhesion of the film.

How paints dry

A paint should remain fluid if left in the can and should dry quickly when spread on a surface. The fact that sometimes paint does not dry readily suggests that drying is not just a case of simple evaporation of the liquid.

The drying of a paint can involve one or more of the following processes:

(a)   *Evaporation*
Some paints dry wholly or partially by evaporation, e.g. nitro-cellulose lacquers and emulsion paints. The polymer (large mole-cule formed by polymerisation) is, in fact, fully formed in the can and when the solvent has evaporated it becomes hard and not sticky. Emulsion paints harden mainly by evaporation of the emulsifying liquid, which is normally water. These paints dry quickly and are easily handled and stored.

(b)   *Drying by reaction between the paint and air*
Oxygen and water vapour are active ingredients of the air. When the paint is applied, a large surface area is exposed and the solvent evaporates. The oxygen or the water vapour then cause the drying oils and resins to polymerise and cross-link.

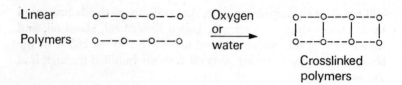

Linear Polymers → Oxygen or water → Crosslinked polymers

A hard tough cross-linked film results which will no longer dissolve in the solvents used in the paint. The chemical reaction with air continues long after the paint is apparently dry to touch. As a result, the paint film goes on changing slowly all the time it is in use, causing the properties of the paint to alter with the passage of time. The drying time depends on temperature and can be very slow in cold weather. Oleoresinous paints harden this way.

(c) *Drying by chemical reaction between constituents of the paint*
To prevent chemical reactions taking place in the can, the active constituents of the paint are separated into two containers and are mixed just before use. Alternatively, chemical reaction may only occur at higher temperatures; this is the case with stoving finish or enamel. It is possible to avoid two packs by the addition of a high proportion of solvent. This slows down the reaction in the can. When the paint is applied and the solvent evaporates, the reaction proceeds more rapidly.

This type of paint dries fairly quickly, the drying time being dependent on the temperature. Examples are polyurethane and polyester finishes.

Types of paint and uses

It is appreciated that there is more than one way of classifying paints and that each method will have its advantages and disadvantages. The classification used here depends upon the function of the paint, namely whether the paint is used as a primer, undercoat, or finishing coat. A section for paints and other materials which do not belong clearly to one or other of the previous groups is also included.

Primers

The first coat of paint on any surface, the primer, should adhere firmly to that surface and provide a suitable base to receive and hold the next coat. It follows that the right primer must be used for different materials.

Metal primers
For metal surfaces liable to corrosion, the primer needs to be particularly adhesive and should contain pigments which inhibit the corrosion. Read the chapter on corrosion of metals for further details.

Wood primers
A wood primer should provide an adherent flexible coating which is moisture-resistant and which limits the absorption of further coats of paint. Examples of wood primer are:
(a) *Lead-based* These consist of a mixture of red and white lead pigments in a linseed oil medium. Resins are normally added to give better spreading, levelling, and drying properties.
(b) *Leadless* These do not protect as well as the lead-based primers and should only be used for interior work. They consist of leadless white pigment, often with colouring matter in an oleoresinous or alkyd medium.
(c) *Aluminium* Fine flakes of aluminium metal are dispersed in an oleoresinous or alkyd vehicle. It provides a high degree of resistance to the passage of moisture and is very suited for sealing the end grain of wood.
(d) *Cement and plaster* The function of the primer is to provide a good foundation for the subsequent paint and prevent chemical attack on the paint by alkaline constituents of the plaster. The primer usually consists of a lime-resistant pigment in a vehicle of tung oil and phenolic resin or tung oil and coumarone resin and chlorinated rubbers.

It should be said that if the surface is heated, the primer should retain its adhesion and should not become brittle.

Undercoat

The functions of an undercoat are:
(a) To cover the primed surface.
(b) To provide a fresh surface of uniform absorbing power and of a similar colour to that of the finishing coat.
(c) To help build up a sufficiently think layer of paint to protect the painted surface.

The type of undercoat used is decided by the durability and weather-resistance required and the contrast in colour between the original surface and the paint finish. Thus, undercoats used for exterior finish need to be more flexible, durable, and water-resistant than for interior use, and should be of greater thickness.

Finishing paints

The final coat should provide the required colour, texture, and protection against weathering, etc. The following abbreviated list includes some of the materials now available. It should be realised, however, that detailed characteristics and uses are given in the manufacturer's literature.

Oil gloss
These are the older paints which consist of a mixture of pigments with extenders ground in linseed oil and thinned with white spirit or turpentine.

Hard gloss (oleoresinous)
These are based partly or wholly on mixtures of drying oils and natural or synthetic resins. The resins are dispersed finely throughout the vehicle and these provide opacity, flow, gloss, and colour. They can be classified depending on the type of resin used. Thus, there are, for example:

Alkyd resin based paints
As already indicated these can be further classified depending on the type of alkyd resin used. They have good durability, but often show loss of adhesion in wet conditions.

Polyurethane resin based paints
When used in conjunction with drying oils, paints are produced which give a high resistance to water, alkali, and atmospheric attack.
Other resins used include phenolics and epoxy esters.

Bituminous paints
Essentially, these are solutions of natural bitumens and bitumen residues from petroleum refining, in hydrocarbon solvents. They are black, and offer high resistance to moisture and chemical attack. They are, however, sensitive to ultra-violet light.

Rubber paints

These are used to provide high resistance to chemical attack and to water. They are solutions of chlorinated natural rubber with added plasticisers and chemically resistant pigments. Isomerised rubber paints have similar properties.

Organosols

These comprise dispersions of finely divided resin particles, essentially polyvinyl chloride, in a volatile organic liquid. The paint is applied and then heat treated to form a continuous film. These can be highly resistant to chemical attack.

Plastisols

These use polyvinyl chloride resin in a mixture of plasticisers. Compared with the organosols, they do not contain a volatile solvent and the shrinkage of the resultant paint film is consequently smaller.

Water paints

All paints under this heading use water for the bulk of the vehicle:

Distemper

Consists of powdered chalk or similar material tinted with colouring pigment and mixed with a solution of glue or a similar binder. It finds use as a temporary decoration, because it is easily washed off.

Oil-bound water paint

This consists of white and tinting pigments plus extenders dispersed in emulsions of drying oils or oil varnishes in water containing glue. A flat finish results which develops good resistance to washing within a few weeks of application.

Emulsion paints

These are sometimes called plastic emulsion paints. They consist of dispersions of synthetic resin particles in water with suitable colloids and dispersing agents. Suitable pigments provide the required colour and opacity. After application, the resin particles form a fairly continuous low gloss film. Emulsion paints may be used on newly plastered walls, cement, etc. Some types may also be used for exterior surfaces. (A colloid is a substance which contains particles in suspension which are too small to be detected with the naked eye. Colloids resist settling out for a long period of time.)

Vinyl water paints

These are intermediate in properties between emulsion paints and oil-bound water paints. They are less washable than emulsion paints but are easy to apply and dry rapidly. They use a polyvinyl acetate copolymer as binder, and have a higher pigment content than emulsion paints.

Textured paints

These are powders which consist of whiting, gypsum, asbestos, mica, coloured pigments, and glue or some other binding agent. They have to be mixed with water before use. Alternatively they may be stiff paste supplied with a suitable vehicle. They are used to provide a variety of interior textured finishes. Manipulation is achieved with brushes, combs, knives, etc.

Chemical-reaction-type coatings

For reasons already stated, a large number of these are two-pack materials. Many of them are now available and they include epoxy resin, phenolic resin, polyurethane, polyester, and neoprene finishes.

The epoxy resin type was first developed for use with anhydride and amino curing agents. The cure depends on temperature and time and the formula can be altered to develop resistance to water and chemical attack.

Thixotropic paints

Thixotropic or non-drip paints are now very widely used. They have a jelly-like consistency and high viscosity when at rest, for example in the can. When agitated or subjected to any shearing action, the viscosity is greatly reduced and they flow quite readily. If such a paint is then left for a period of time its viscosity rises and its jelly-like properties return.

This property, termed 'thixotropy', is achieved by the use of additives such as a small amount of silica, metallic silicates, or certain organic polymers such as hydrogenated castor oil, which are known as resinous thickeners.

Miscellaneous painting materials

Knotting compound
This is a dispersion of lac or some other natural or synthetic resin in a suitable solvent. On drying, it should form a film which is impervious both to the solvent in the knots and in the paint.

Wood stain
This is an oil, water, or alcohol (spirit) medium containing soluble dyes or transparent pigments. The stain should produce an overall penetration of the wood surface.

Cement paint
This consists of white or coloured Portland cement plus accelerators, extenders, and waterproofing agents. It is supplied as a powder which is mixed with water and applied.

Clear finishes for wood
These are of two types:
(a)   Coatings which form a clear relatively impervious coating on the wood. These include conventional varnishes plus a widening range of synthetic coatings.
(b)   Coatings based on a drying wax or oil. These penetrate into the wood and produce soft films which retain dirt. They should therefore not be used to excess.

Cementiferous paints
These are paints used to inhibit the corrosion of steel. They contain large quantities of finely divided metallic zinc in an inorganic binder. When dry, they form a hard cement-like film. They are normally supplied as a two-pack system.

Painting defects

Defects in paintwork are generally due to lack of preparation of the surface and/or not following the manufacturer's instructions. Having said that, the following is a list of fairly common defects, their causes, and remedies.

Bittiness
Bits of dirt from the atmosphere or the brushes, or skin

from the paint, harden in the paint film, producing a rough unsightly finish (Fig 33). It can be cured by rubbing down and repainting.

Bleeding
This is the staining of a coat of paint or varnish by soluble substances diffusing to the surface from undercoatings. It is caused by
painting directly over materials such as bituminous paint, which
have not been properly sealed. It can be cured by complete removal
of the paint or by sealing and then repainting.

Blistering
This is caused by moisture or solvent which is trapped in the substrate below the paint film producing differential swelling of the
film (Fig 34). It can arise through painting in the direct heat of the
sun which causes the surface skin to form quickly and traps the
solvent. Painting over a surface with a high moisture content can
also produce blistering. If the blistering is severe, the paint should
be stripped off and repainted. Scattered blisters can be burst,
flattened, and repainted.

Cissing
This is the shrinkage or drawing away of a new coating from large
or small areas of the work (Fig 35). It occurs most readily with
varnish and is caused mainly by contamination of the surface with
grease, oil, polish, or exudation from an oily undercoat. It can be
avoided by making sure that all surfaces to be painted are clean.
A cissed surface is cured by rubbing down and repainting.

Blooming
This describes a reduction of gloss on certain finishes owing to the
formation of a mist or haze. It is caused by moisture or atmospheric
contamination and it can be cured by polishing with a soft cloth or
by gentle washing. To prevent it, painting should be carried out
in warm dry conditions.

Chalking
This is the formation of a fine powder on the paint film. It may be
due to a paint deficient in binder or a normal paint applied over a
porous surface. To prevent it, seal surfaces before painting, use
good quality paints, and do not mix undercoat into the finish.

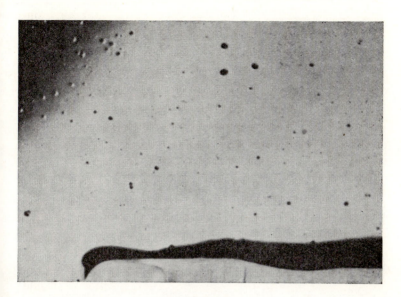

Fig 33 Bittiness

Fig 34 Blistering

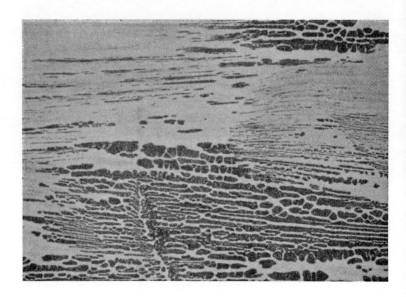

Fig 35 Cissing

Fig 36 Crazing or cracking

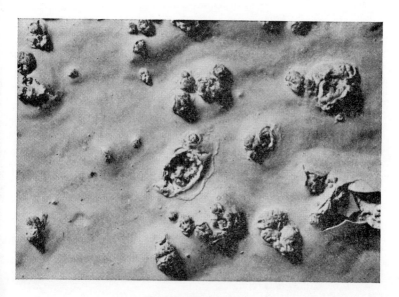

Fig 37 Efflorescence

Fig 38 Flaking

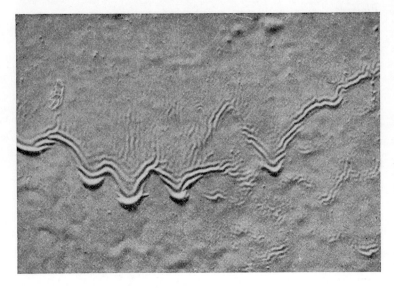

Fig 39 Running or sagging

Light chalking may be cured by wiping; if heavy, the paint may need removing.

Crazing or cracking
This is the irregular cracking of a surface coating and is due to the age of the coat or the application of a hard drying paint over an oily undercoat, an undercoat which has not fully dried, or a bituminous surface (Fig 36). It may be prevented by ensuring that all paints to be overcoated are fully hardened. Rubbing down and recoating will cure it, if slight; if severe, stripping down is necessary.

Efflorescence
This occurs on brick, plaster, and cement and is caused by soluble salts coming to the surface in solution and being deposited on the surface (Fig 37). Paint applied to these surfaces will be disfigured and possibly disrupted. If it appears on the paint film without breaking it, it may be wiped off. Should the film be broken, the whole area should be stripped, wiped, and left until efflorescence stops. It should then be safe to decorate.

Flaking

The paint lifts away or peels back from a surface usually from a split or joint in the film (Fig 38). It can be due to:

(a)   The use of an unsuitable primer or undercoat.

(b)   Application over powdery or chalky old paint or dirty greasy surfaces.

(c)   Shrinkage or expansion of the surface.

It is prevented by painting sound clean surfaces with the recommended primer. It is cured by stripping completely and repainting.

Lifting

This is the disturbance of the previous coat when the new paint is applied. The top coat then tends to wrinkle or roll up under the brush. It can result if the previous coat is not thoroughly dry or if the old paint is not resistant to the solvents in the new paint. To prevent it, make sure that the surface paint is dry and older suspect coatings are stripped off. If lifting has occurred, it can only be cured by stripping the surface down.

Running or sagging

In general, this is caused by uneven application of the paint (Fig 39). The effect is increased by putting paint on a painted edge which has not set properly. It is also noticeable where too much paint has been applied to a riveted or moulded surface. It can be prevented by taking greater care in the painting of moulds, etc. It can be cured by allowing the paint to dry thoroughly, then rubbing down and repainting.

The testing of paint

The nature of the paint is determined by the properties of the pigments used and the properties of the vehicle. It is therefore necessary to determine the properties of the constituents and the properties of the paint itself once it has been produced. British Standard 3900 (B.S. 3900) gives a full list of the properties to be tested and the procedures to be followed. Only a selection of those tests will be mentioned here.

Tests on pigments

(a) *Tinting strength*
This measures the amount of pigment which has to be mixed with a white pigment in order to produce a given colour.

(b) *Lightfastness*
This measures the ability of the pigment to retain its colour when exposed to light.

(c) *Bleeding characteristics*
Determination of the solubility of pigments in different solvents. Thus, the appearance of the colour of an older undercoat through the top coat (bleeding) can be avoided.

(d) *Hiding power*
This is expressed as the area in square metres covered by one kilogramme of pigment which has been dispersed in a paint and applied so that it just hides any previous colour.

(e) *Particle size*
The maximum scattering of light is produced by particles of diameter approximately half the wavelength of light, i.e. of diameter between 2 and 4×10^{-7} m. Particle size thus influences the evenness of the colour of a paint film.

(f) *Particle shape*
The shape of the pigment, viz. spherical, cubic, rounded, irregular, needle-like, or plate-like, determines the packing of the pigment and hence the covering power of the paint. Waterproof paints include plate-like pigments which overlap like tiles on a roof; aluminium and mica pigments are used.

(g) *Thermal stability*
The temperature at which a pigment decomposes or melts can be vital when deciding whether or not to use a particular pigment in a heat-resistant paint.

Testing of solvents

The solvent is the liquid or mixture of liquids used in the paint. The following are important characteristics of the solvents:

Boiling point and evaporation rate
The flow of paint on a vertical surface is controlled by the evaporation of the solvent. For most purposes, the boiling points are a good guide to the evaporation rate. Liquids of low boiling point evaporate more readily at room temperature than liquids of high boiling point. Low boiling solvents are used in spraying paints and high boilers to give flow. Thus, boiling point and boiling range are important.

Flash point
The flash point is the lowest temperature at which enough vapour is given off to form a mixture of air and vapour above the liquid which can be ignited by a spark or flame under specified conditions. Factory regulations and regulations governing the transport of materials require precise knowledge of flash points. The Abel flash-point apparatus is commercially available.

Toxicity and smell
Some liquids have a cumulative poisonous effect and others can be harmful above certain concentrations. The smell of a solvent alone could restrict its use.

Tests on paint

Viscosity measurements
The viscosity determines the flow properties of the paint. One apparatus used to measure viscosity uses a constant speed electric motor to drive a stirrer which is immersed in the paint. The paint container stands on a turntable with a torque measuring attachment. The torque or opposition to the rotation of the stirrer is directly related to the viscosity and the instrument is calibrated directly in poises (viscosity units).

Density measurements
Measuring the mass per unit volume of the paint.

Hardness determination
The paint film is allowed to harden and subjected to a sudden force at a point using an impact testing machine.

Abrasion resistance
The force or effort required to scratch the hardened paint film is
determined.

Hard drying time
A loaded mechanical thumb is forced around on the surface of the
test panel and the drying time is determined.

Opacity
Opacity, or the hiding power of a paint film, is measured by an
instrument which is called a cryptometer.

Adhesion, flexibility, and resistance to weathering are also
measured. Exposure tests are carried out on standard size panels
which are set up at 45° to the horizontal facing south. They are
examined over a period of years. The resistance of the paint to salt
water may be determined using a salt spray apparatus.

EXPERIMENT 32
To determine the density of the paint

Apparatus
Standard density cups or 100 ml measuring cylinders, beakers,
spatula, paint solvent, ready-mixed paints, chemical balance.

Procedure
(a) Weigh a measuring cylinder empty.
(b) Stir the paint thoroughly, then pour into a beaker.
(c) Fill the measuring cylinder up to the 100 ml mark and weigh.
(d) Express the density in kg/l.
(e) Repeat with the other materials, remembering to clean the
apparatus thoroughly between experiments.

EXPERIMENT 33
Determination of spreading power

Apparatus
Primed panels of area 0·1 m², beakers, 25 mm paint brushes,
undercoat, finishing paint.

Procedure for undercoat
(a) Mix the paint thoroughly and pour into a small beaker.

(b)   Weigh the beaker plus paint and brush.
(c)   Apply sufficient undercoat to cover the panel.
(d)   Re-weigh the beaker, paint, and brush.

Top coat
Allow the undercoat to dry, then repeat the above procedure with
the top coat.

$$\text{Spreading power} = \text{Area covered} \times \frac{\text{Density of paint}}{\text{Mass of paint used}}$$

Questions
1   Name the different classes of pigment used.
2   What are the functions of a pigment?
3   What is meant by the term 'vehicle' in relation to paints?
4   What is a resin?
5   List the properties which resins confer on paints.
6   What is the main advantage of synthetic resins over natural
resins?
7   Give a brief description of the hardening processes of paints.
8   Describe the functions of (a) the primer,
                              (b) the undercoat,
                              (c) the finishing coat.
9   Why does the paint film 'deteriorate' with the passage of time?
10 What is a thixotropic paint?

Nine The Corrosion of Metals

Corrosion is the name given to the degradation of a metal by its chemical combination with a non-metal such as oxygen or sulphur. Generally this means a return of the metal to the form in which it existed as a naturally occurring ore with a complete loss of metallic properties. The metals which are least prone to corrosion are those which are most readily obtained from the ore, e.g. gold and silver. These metals, which are obtained with the greatest difficulty, tend to revert to their natural state most readily, e.g. iron. Corrosion can only be explained as an electro-chemical process, so it is first necessary to have some idea of the make-up of atoms and the process of electrical conduction through liquids.

Simple picture of the atom—ionisation
All elements, including the metals, consist of atoms, the atom being defined as the smallest part of a substance which takes part in a chemical reaction. An atom can be considered to contain a positively charged massive centre or nucleus which is surrounded by a cloud of negative charge (Fig 40). The electron is the unit of negative charge. In a neutral atom the positive charge on the nucleus is equal to the number of electrons around the nucleus.

Ionisation

When a metal atom loses one or more electrons the process is called 'ionisation'. A positively charged atom or positive ion is formed plus a free electron,

i.e. $\qquad M \xrightarrow{\text{Ionisation}} M^{n+} + n.e^-$

Metal $\qquad\qquad$ Metal
$\qquad\qquad\qquad\qquad$ ion

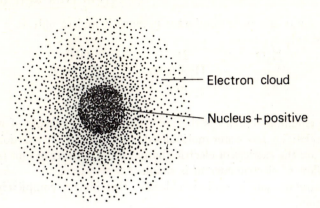

Fig 40

where n is a whole number and e^- represents the electron.

Non-metals ionise when the atom gains an electron,

i.e. $X + ne^- \longrightarrow X^{n-}$

Non-Metal   Non-Metallic

atom   ion

Conduction of electric current through liquids

Pure water does not conduct electricity. This is demonstrated by placing some pure distilled water in a beaker and immersing two copper plates in the water (Fig 41). The plates are then connected to a battery via a switch and light bulb as shown. When the switch is closed no current flows but if a little common salt is added the bulb lights up showing that current is flowing. The explanation is

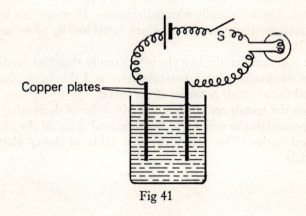

Copper plates

Fig 41

that water is a poor conductor because it ionises only to a very slight extent,

i.e. H_2O $\rightleftharpoons$ H^+ $+$ OH^-
            Water   Hydrogen   Hydroxyl
                               ion   ion

The equation represents the ionisation of one molecule of water but relatively few water molecules do in fact ionise. In solution the ions are the carriers of electric charge; if the number of ions is low the flow of electric current is small.

Common salt is sodium chloride, this ionises completely on solution,

i.e. $NaCl$ $\longrightarrow$ Na^+ $+$ Cl^-
        Sodium chloride   Sodium   Chloride
                                             ion   ion

The solution now contains a large number of charge carriers or ions so it is a good conductor of electricity.

A solution which conducts electricity is called an 'electrolyte'. Acids, alkalis, and many other solutions ionise to a greater extent than water and therefore act as good electrolytes.

Electrode potential—the electrochemical series

When a metal is placed in contact with a solution of its own ions metallic ions are transferred between the metal and the solution and a potential difference is created between the metal and the solution. The size of the p.d. depends on the nature of the metal and the concentration of the electrolyte. To measure this p.d. a similar electrode of known potential, i.e. a reference electrode, is required. The universally adopted reference electrode is a hydrogen electrode, the potential of which is arbitrarily taken as zero (Fig 42).

The p.d. between the two electrodes under standard conditions is then measured using a potentiometer and the value obtained is called the *electrode potential* of the metal.

When the metals are written down in order of decreasing electrode potentials the series which is obtained is called the electrochemical series. The following is a table of some electrode potentials.

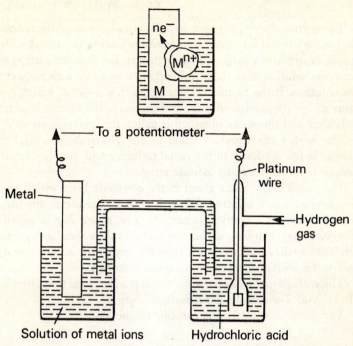

Fig 42

	Ion in solution/Metal	Electrode potential (volts)
Lithium	Li⁺/Li	−3·04 Anodic
Potassium	K⁺/K	−2·92
Calcium	Ca²⁺/Ca	−2·87
Sodium	Na⁺/Na	−2·71
Magnesium	Mg²⁺/Mg	−2·37
Aluminium	Al³⁺/Al	−1·66
Manganese	Mn²⁺/Mn	−1·18
Zinc	Zn²⁺/Zn	−0·76
Chromium	Cr²⁺/Cr	−0·71
Iron	Fe²⁺/Fe	−0·44
Nickel	Ni²⁺/Ni	−0·236
Tin	Sn²⁺/Sn	−0·136
Lead	Pb²⁺/Pb	−0·126
Hydrogen	H⁺/½H₂	0
Copper	Cu²⁺/Cu	+0·34
Silver	Ag⁺/Ag	+0·80
Gold	Au³⁺/Au	+1·42 Cathodic
Chlorine	Cl⁻/½Cl₂	+1·36

G

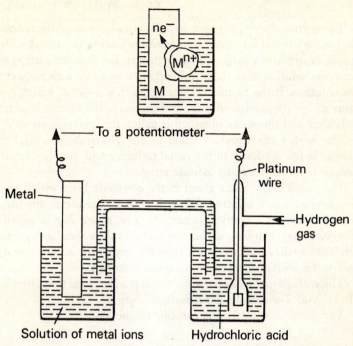

Fig 42

	Ion in solution/Metal	Electrode potential (volts)
Lithium	Li^+/Li	-3.04 Anodic
Potassium	K^+/K	-2.92
Calcium	Ca^{2+}/Ca	-2.87
Sodium	Na^+/Na	-2.71
Magnesium	Mg^{2+}/Mg	-2.37
Aluminium	Al^{3+}/Al	-1.66
Manganese	Mn^{2+}/Mn	-1.18
Zinc	Zn^{2+}/Zn	-0.76
Chromium	Cr^{2+}/Cr	-0.71
Iron	Fe^{2+}/Fe	-0.44
Nickel	Ni^{2+}/Ni	-0.236
Tin	Sn^{2+}/Sn	-0.136
Lead	Pb^{2+}/Pb	-0.126
Hydrogen	$H^+/\frac{1}{2}H_2$	0
Copper	Cu^{2+}/Cu	$+0.34$
Silver	Ag^+/Ag	$+0.80$
Gold	Au^{3+}/Au	$+1.42$ Cathodic
Chlorine	$Cl^-/\frac{1}{2}Cl_2$	$+1.36$

G

The metals above hydrogen in this series have negative potentials because metal ions pass into solution leaving the metal with a resultant negative charge. Below hydrogen the ions are plating out from the solution on to the metal making it positive with respect to the solution. If the metals are immersed in sea-water, which contains a high proportion of dissolved minerals, the electrode potentials alter and the series obtained is called the 'galvanic series'.

The bigger the negative value of the electrode potential the greater is the tendency of the metal to ionise and dissolve, i.e. the greater is its tendency to corrode away.

The *anode* is the name given to the electrode from which positive ions pass into solution. When ions leave the metal, the metal always becomes negatively charged. *The anode has a negative charge*. Thus it can be said that any metal in the series has a greater tendency to dissolve or corrode than the one below it. The metal is said to be *anodic* with respect to those below it.

The *cathode* is the electrode to which positive ions flow from the electrolyte. *The cathode has a positive charge*.

Any metal in the series is *cathodic* to one above it.

The chemical cell

When two different metals are immersed in an electrolyte, e.g. zinc and copper in dilute sulphuric acid, and the metals are connected externally through an ammeter, current flows from one electrode to the other (Fig 43). The metal which is anodic dissolves or corrodes and the electrons remaining on the anode travel through the connecting wire to the cathode where they neutralise positive ions.

In the case of zinc and copper, the zinc is anodic and corrodes away whilst positive ions plate out on the copper which is cathodic. At the same time hydrogen gas is evolved at the cathode.

Conditions under which a metal corrodes

It is now possible to make a general statement about the conditions under which a metal corrodes.

Corrosion is the result of simultaneous processes at an anode and cathode. The anode and cathode must be in metallic connection and must be immersed in an electrolyte. The rate of corrosion is decided by the size of the corrosion current and one factor

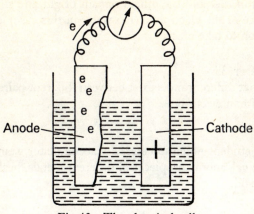

Fig 43   The chemical cell

which decides this is the potential difference between anode and cathode. The anode and cathode do not have to be different metals but can be different areas of the same metal as will be seen when the corrosion of iron is dealt with.

EXPERIMENT 34
To show the presence of an electrolyte is essential to corrosion.

Apparatus
Two test tubes, nails, calcium chloride, cotton wool.

Procedure
Place a few nails in a test tube and cover with tap water. Place a few similar nails in a test tube, plug with cotton wool, and place a layer of anhydrous calcium chloride on the cotton wool as shown in Fig 44. Leave for about one week. The nails in the water will

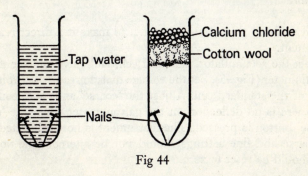

Fig 44

be covered in rust and the others remain bright and shiny. Thus the presence of the electrolyte (tap water in this case) is essential for corrosion to take place.

EXPERIMENT 35
To show that different p.d.s exist between different pairs of metals immersed in an electrolyte.

Apparatus
Pye model student potentiometer, 2 v accumulator, standard cell, centre zero galvanometer, pieces of different metals, beaker, dilute sulphuric acid.

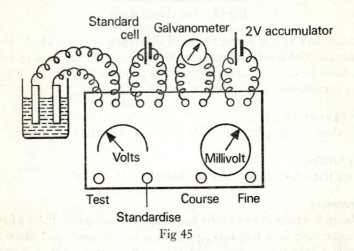

Fig 45

Procedure

A *To calibrate the potentiometer,* i.e. to make it a direct reading instrument.
Connect the accumulator, standard cell, and galvanometer to the potentiometer (Fig 45). Set the voltage dials to read 1·018 volt, the e.m.f. of the standard cell. Adjust the 'coarse' and 'fine' controls until there is no deflection on the galvanometer when the 'standardise' button is pressed. The instrument is now calibrated and the coarse and fine settings should not be altered. The voltage dials should be reset to zero.

B *To measure the unknown p.d.s*

Partly immerse a pair of metals in the electrolyte and connect them to the test terminals. Adjust the voltage controls until when the test button is pressed the galvanometer deflection is zero. The p.d. between the metals is read directly off the instrument. Repeat for different pairs of metals. Record your observations and relate them to the electrode potential series. The experiment may be repeated using a different electrolyte, e.g. a solution of common salt.

The rusting or corrosion of iron

If a piece of iron or mild steel is placed in tap water it corrodes fairly quickly. If the iron were examined closely, it would be found to be covered with tiny anode and cathode areas. The anodes would be dissolving. Fig 46 illustrates one such corrosion cell, greatly magnified. Many factors contribute to the production of anodic and cathodic areas, among them:

(a)   Any inclusions (impurities) in the metal.
(b)   Imperfections in the surface of the metal.
(c)   The orientations of the grains.
(d)   Localised stresses in the metal.
(e)   Variations in the surroundings of the metal.

   Corrosion proceeds as follows. Ferrous ions Fe^{2+} leave the anode. The released electrons travel through the metal to the cathode where they neutralise hydrogen ions to form hydrogen gas. The gas collects on the cathode which effectively insulates the

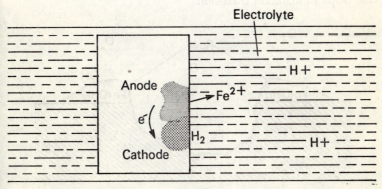

Fig 46

cathode from the electrolyte if it is not disturbed. The cathode is then said to be polarised.

Reactions : At the anode $Fe \rightarrow Fe^{2+} + 2e^-$
At the cathode $2H^+ + 2e^- \rightarrow H_2$ (gas)

If the electrolyte is acidic the hydrogen gas bubbles off the cathode and the metal corrodes rapidly away. Thus a piece of mild steel dissolves rapidly in hydrochloric acid. If the electrolyte is neutral or slightly alkaline, the action of dissolved oxygen in the corrosive process becomes very important.

Effect of dissolved oxygen
Oxygen dissolved in the electrolyte combines with the hydrogen which forms on the cathode producing water.

$$\text{Possible reaction}\quad 2H_2 + O_2 \rightarrow 2H_2O$$

The insulating layer is thus removed from the cathode which is said to be depolarised. More of the anode now dissolves and electrons travel through the metal to neutralise more hydrogen ions. In short the iron corrodes in neutral or slightly alkaline solutions only in the presence of dissolved oxygen.

Rust is the familiar reddish brown covering formed when iron corrodes. It is formed when the ferrous ions Fe^{2+} and hydroxyl ions OH^- come into contact whilst moving in opposite directions through the electrolyte (Fig 47). They react to form ferrous hydroxide $Fe(OH)_2$ [$Fe^{2+} + 2OH^- \rightarrow Fe(OH)_2$]. This is quickly oxidised by dissolved oxygen to ferric hydroxide $Fe(OH)_3$ which is rust. Rust is porous so that moisture is retained beneath it; this helps to maintain corrosion.

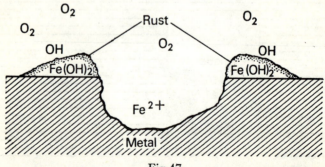

Fig 47

EXPERIMENT 36
To show the presence of anodic and cathodic areas on a piece of
mild steel.

Apparatus
Mild steel plate, suitable size 75 mm × 75 mm × 6 mm.
Corrosion indicator solution.

The *corrosion indicator solution* consists of 0·5 ml of 1% alco-
holic phenol phthalein and 3 ml of freshly prepared 1% potassium
ferricyanide added to 100 ml of tap water containing 0·6 g of
sodium chloride.

Procedure
If necessary, first clear the steel plate of rust. Rub the surface to
be treated with coarse emery cloth and place a drop of the indicator solution on the surface of the metal.

Small pink and blue spots appear under the drop and persist
until all the oxygen dissolved in the indicator is used up.

This is known as the primary corrosion. The cathodes are indicated by the pink spots. At the cathodes an excess of hydroxyl ions
OH^- is formed and causes the phenolphthalein to turn pink (Fig
48).

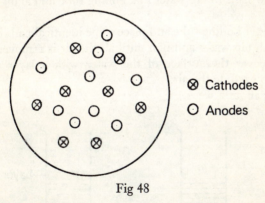

Fig 48

The anodes are indicated by the blue spots where an excess of
ferrous ions Fe^{2+} causes the potassium ferricyanide to turn blue.

Secondary corrosion
In time the formation of OH^- at the centre of the drop ceases due
to the lack of oxygen. At the edge of the drop there is a high oxy-

gen concentration so this becomes cathodic and turns pink. The centre turns blue, i.e. it becomes anodic, owing to the concentration of Fe^{2+}. In time the ferrous ions react with the water to form a brown ring of rust between the anodic and cathodic areas (Fig 49).

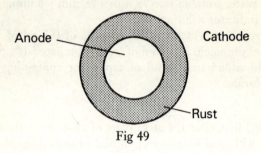

Anode

Cathode

Rust

Fig 49

EXPERIMENT 37
To show the presence of oxygen is esssential for iron to corrode in neutral solution.

Procedure
Place a few nails in tap water in a boiling tube and allow to stand (Fig 50).
Take another boiling tube and place some identical nails in it.
Cover with tap water and boil until all the air is removed. Whilst still hot cover the surface of the water with a layer of liquid paraffin to exclude the air.
Leave for about a week.

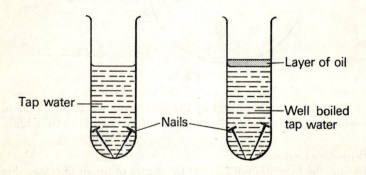

Tap water

Nails

Layer of oil

Well boiled tap water

Fig 50

The nails in the tap water should rust freely whilst the others remain rust-free.

The lack of oxygen in tube (2) prevents the cathodic reactions taking place, i.e. the corrosion cells polarise thereby preventing corrosion and rust formation.

Note : If rust does form in tube (2) it is due to the presence of dissolved oxygen.

The effect of the presence of other metals on the corrosion of iron.
Metals which are anodic to iron
Any metal higher in the electrode potential series than iron will be anodic to the iron if in contact with it in the presence of an electrolyte. This anodic metal will corrode in preference to the iron. The bigger the p.d. between the metals the more rapidly the metal corrodes, e.g. zinc is anodic to iron and will form a *sacrificial anode* when in contact with iron (Fig 51). This is amplified later in the section on protection against corrosion.

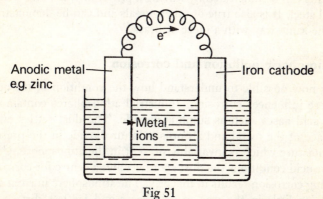

Fig 51

Metals which are cathodic to iron
Iron is anodic with respect to any metal below it in the electrode potential series; alternatively it can be said that metals below iron are cathodic with respect to it. Thus a piece of iron in contact with a piece of tin, for example, will suffer accelerated corrosion as long as the conditions for creating a simple cell exist.

The effect of differential oxygen concentration
If the supply of oxygen to any part of an iron surface is limited the iron which is oxygen starved becomes anodic and corrodes (Fig 52). The supply of oxygen is limited by the presence of dust or dirt

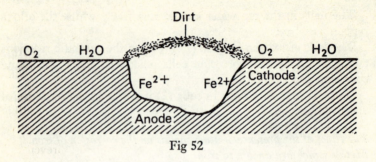

Fig 52

or by the presence of any other material on the surface. The chemical reactions at anode and cathode are as previously described. This accounts for rust creeping under damaged paint films and for corrosion just below the water-line in a steel tank.

Below the water-line the oxygen concentration will be lower causing it to become anodic and corrode.

This can be illustrated by leaving a pile of oily dirt on a piece of mild steel. It is also true of other metals and can be demonstrated in the same way with a piece of brass.

Atmospheric pollution and corrosion

It is now possible to understand how the condition of the atmosphere influences corrosion. Industrial atmospheres contain dust and acid gases such as sulphur dioxide. The dust settles on the surface of the metal and absorbs sulphur dioxide in the presence of moisture which is invariably present in the atmosphere. Under these acid conditions the dust spot becomes a centre for corrosion. Pitting corrosion results in this way. The atmosphere in rural areas contains little in the way of acid gases and far less dirt so that corrosion is not so severe. If the atmosphere is very dry the corrosion is reduced because there is insufficient electrolyte to establish the corrosion cells.

Anaerobic corrosion

Anaerobic bacteria are those which do not require oxygen in order to exist. These bacteria are found particularly in waterlogged clay, and ironwork buried in such clay is found to corrode rapidly. The bacteria catalyse, i.e. speed up the reaction between hydrogen ions (from the electrolyte) and say calcium sulphate present in the soil. This effectively depolarises the cathodic areas and causes the corrosion to proceed rapidly. Once started the corrosion spreads rapidly due to the multiplication of the bacteria. Black ferrous sulphide is the main corrosion product.

Anodic reaction	$Fe \longrightarrow Fe^{2+} + 2e^-$
Electrolyte	$H_2O \longrightarrow H^+ + OH^-$
Possible bacteria	$SO^{2-}_4 + 8H^+ + 8e^- \longrightarrow S^{2-} + 4\,H_2O$
Catalysed reaction	sulphate ion sulphide ion
	$Fe^{2+} + S^{2-} \longrightarrow FeS$
	Black Ferrous sulphide

A *catalyst* is a substance (produced in this case by the bacteria) which increases the rate at which a chemical reaction occurs. The catalyst is unchanged at the end of the reaction.

This type of corrosion may be prevented by covering the iron-work, e.g. pipes, with thick layers of bitumen before burial. It can also be stopped by supplying oxygen, i.e. aerating the affected parts. This kills off the bacteria.

Stray current corrosion

This can be important if a buried metal structure is in close proximity to a high voltage source of direct current, e.g. electric railway lines which are not too well insulated from earth. The current will flow through the structure causing the anodic parts of it to corrode. It is controlled by reducing the flow of current through the earth and through the structure. In some cases the anode is made purposely expendable.

The corrosion of some commonly used non-ferrous metals

Aluminium

The surface of pure aluminium metal oxidises rapidly to form a minutely thin film of aluminium oxide which is very coherent and adheres closely to the aluminium metal. This oxide layer protects the metal against further corrosion. It does, however, corrode in all kinds of salt solutions because aluminium chloride and sulphate for example do not form coherent films on the surface of the aluminium. This accounts for the condition of unpainted aluminium container vehicles, particularly during the winter months.

Copper

A new copper surface appears to be reddish in colour. However it rapidly corrodes in air becoming covered with an adherent brown film of copper sulphide or oxide. On prolonged exposure a green film of basic copper carbonate, sulphate, or chloride is formed. The basic salt is a complex of either of the above with cupric hydroxide $Cu(OH)_2$. These films are coherent and protect the copper against further corrosion. Copper however does corrode in the presence of organic acids and salts. Copper nozzles on pipes corrode rapidly because the relative motion of the liquid layers flowing over its surface causes corrosion cells to be set up which result in rapid wear.

Lead

Lead is normally protected by a coherent film of oxide. The formation of the oxide can be observed on the surface of a freshly cut piece of lead which rapidly loses its lustre. It does however corrode if buried in loams or acid soils. The presence of alkaline substances also assists the corrosion of lead. In the section dealing with water it is stated that the formation of an impervious film of basic lead carbonate prevents further corrosion of the lead pipe, when used in hard-water areas.

Zinc

Zinc metal is reasonably stable in air. being covered with a coherent film of oxides, hydroxides, and carbonates which protect it. Zinc is anodic to iron and it is used to protect iron structures.

The prevention or limitation of corrosion

The general methods of preventing corrosion are outlined in this section, although it should be remembered that each example of corrosion is a particular case. It has already been pointed out that the electrode potential of a metal alters under non-standard conditions, and this has to be remembered in dealing with practical problems. Economic factors largely determine the treatment to be adopted and it may even be cheaper to allow the metal to corrode and to replace it when necessary.

The methods of preventing corrosion will now be considered under the following headings:

(a)   Production of corrosion-resistant alloys and careful structural design.

(b)   Cathodic protection.

 (i) Use of sacrificial anodes.

 (ii) Impressed current.

(iii) Use of anodic metal coatings.
(c) Methods which depend on the exclusion of air and moisture.
 (i) Application of coatings of paints, rubber, plastics, metals.
 (ii) Chemical treatment of metals to form stable passive surface films.

Corrosion-resistant alloys and careful structural design
These alloys are stable because a very coherent surface film of a compound of the metal, usually the oxide, forms on the surface. This passive film inhibits further corrosion. Thus metals such as chromium and aluminium are stable in many environments because they are protected by an air-formed oxide film. Stainless steel contains 18% chromium and 8% nickel and this is covered with a very stable oxide film. Stainless steel, however, will corrode if the supply of oxygen required to maintain the oxide film is interrupted, e.g. under a rivet the oxygen concentration is low and this area will become anodic and corrode. Rust itself can and does inhibit further corrosion if it is sufficiently compact and impervious. A nickel chromium steel containing a small percentage of copper produces this type of rust; this alloy corrodes very slowly after the formation of the initial layer of rust.

Design
Corrosion can often be reduced or prevented by thoughtful design. Thus tanks or pipelines should be free from obstructions or crevices where water may collect causing corrosion by differential aeration. Where dissimilar metals are in contact the relative areas are important. A small area of anodic metal in contact with cathodic metal will suffer accelerated corrosion, e.g. steel rivets in copper plate corrode very rapidly; the corrosion current per unit area of the anode is high, causing its accelerated corrosion.

Cathodic protection
(a) *Sacrificial anodes*
The principle of the method is illustrated in Fig 53. A metal of greater electrode potential is connected to the corrosion cell with electrodes B and C. B is anodic to C but A is more anodic than B and will therefore corrode preferentially as electrons flow from A to B. This is called cathodic protection because the metal to be protected B is made cathodic with respect to A. This principle is well known and has been in use since the early part of the nine-

teenth century. Thus buried pipelines and steel structures are protected by connecting sacrificial blocks of zinc to these structures.
Blocks of magnesium alloy are also used to protect ships' hulls.

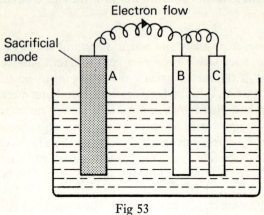

Fig 53

(b) *Impressed or externally applied current protection*
The negative side of a d.c. generator is connected to the electrode
or metal structure to be protected (Fig 54). The anode used can
be of soluble or insoluble metal. Soluble electrodes are not favoured
because they need regular replacement. The method is used for the
protection of ships, submarines, jetties, buoys, and long-distance
pipelines. Platinum or platinum coated anodes are used.

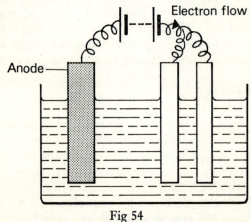

Fig 54

(c) *Use of anodic metal coatings*

A coating of metal which is anodic with respect to the metal to be
protected acts in two ways: (i) it excludes air and moisture, (ii) if
the coating is broken and a corrosion cell is set up then the corrosion of the cathodic metal will be inhibited. Metals commonly
used to protect steel are zinc, aluminium, and chromium coatings;
the steel is first cleaned thoroughly. This is done effectively either
by (i) shot blasting, bombarding the surface of the steel with
abrasive particles, or (ii) flaming, passing an oxy-acetylene flame
over the surface, or (iii) pickling, i.e. dipping the steel in a bath of
hydrochloric or phosphoric acid. The metal coating may then be
applied in one of a number of ways.

Dipping and rolling

The sheet steel is passed through a bath of molten zinc. The metal
then passes through rollers which help to produce an even coating
and squeeze off excess zinc. The coating of steel with zinc is
known as galvanising.

Electrolysis

The steel article to be plated is made the cathode of an electrolytic
cell. Probably the best known example of this is chromium plating.
The steel article, e.g. car bumpers and hub caps, is first coated
with a layer of copper. The copper surface is then nickel plated;
this is highly polished and covered with a minute film of
chromium.

The main disadvantage of this method is that it is difficult to
avoid the formation of minute holes in the coating which allow
corrosion cells to be set up.

Sherardising

This is the application of a metal film by heating the base metal in
the powder of the coating metal. This is used to coat the steel with
a thin layer of zinc by heating it with zinc dust and zinc oxide at
250° to 450°C. If aluminium is used the process is called *calorising*.
The metal object is first cleaned, then heated in a tightly packed
mixture of aluminium and aluminium oxide. Air is excluded.
Chromium also can be applied by a similar method.

Metal spraying

The surface to be sprayed must first be cleaned by grit-blasting or
sand-blasting. Most patterns of spraying pistol contain an oxy-

acetylene blow-pipe flame. The metal may be fed in either as powder or wire and is blown out as atomised droplets in a blast of compressed air. The two metals usually applied by spraying are zinc and aluminium. In this process the sprayed metal adheres to the base metal simply by dovetailing into its surface. In the dipping, sherardising, and calorising processes alloying takes place, i.e. the coating metal dissolves in the base metal forming a solid solution of one metal in another which is called an alloy.

Cladding and tin-plating

Cladding
This consists of making a sandwich of a strong but corrodible material between two thinner layers of the less corrodible but mechanically weaker material. The sandwich is rolled under suitable conditions and the result is a thinner sheet clad with the layer of resistant material. Generally speaking alloying does not take place but the three layers weld together. Cladding is mainly used for light alloys, but nickel-clad steel is widely used in the chemical industry where pure nickel would be too expensive.

The protection given by cladding depends on the exclusion of air and water and so does tin-plating.

Tin plate
This is a well-known and widely used product, which is produced by coating sheet steel with a layer of tin. The process now used is to pass the steel strip continuously through cleaning baths and then through electroplating baths in which tin is plated at a high current density (high current per unit area). The strip is then rinsed and heated in a furnace momentarily to a temperature just above the melting point of tin. This flow brightening process gives the tin plate its lustrous appearance. Hot dipped tin plate, i.e. passing steel through molten tin, produces a greater thickness of tin than electrolytic tin plate and finds use in the manufacture of permanent hardware. Because tin is cathodic to iron the corrosion of the steel will be accelerated if the tin layer is broken in any way.

Rubber, glass, and plastic coatings
Rubber coatings are also used to protect steel containers etc. Sheets of partially vulcanised rubber, i.e. partially hardened rubber, are pressed against carefully sand-blasted and degreased metal surfaces. The process of vulcanisation or hardening is completed by

raising the temperature to 100°C. The rubber adheres more readily if the steel is first brass-plated.

Glass linings and vitreous enamels
Glass-lined steel vessels and glass-lined pipes find applications mainly in chemical works. Thin linings are produced by using vitreous enamels. The material is essentially a boro-silicate glass containing fluorine; this is finely ground, suspended in water or organic solvent, and applied to the surface of the article by dipping or spraying. The article is then dried by warming and its temperature is then raised causing the enamel to melt. This flows together to form a continuous coat.

Polyvinyl chloride (P.V.C.) coatings
Relatively thick coatings of P.V.C. sheet can now be bounded to sheet steel. The coating is adherent and durable and has increased the range of uses of sheet steel.

Chemical treatment to form stable passive surface films
Steel objects can be protected by dipping them in a bath of phosphoric acid. A stable film of iron phosphates is formed which protects the steel against corrosion. The stability of the phosphate layer and hence the effectiveness of the protecting layer is increased by dissolving zinc or manganese phosphates in the bath of hot phosphoric acid. A commercial form of this process, known as 'Parkerising', is a particularly effective way of protecting small objects, e.g. nuts, bolts, washers, camera mechanisms, etc. In commercial phosphating the composition of the phosphating solution is decided to some extent by whether or not the article is to be left unpainted, painted, or covered with oil. Phosphating is often followed by the application of dilute chromic acid or a chromate.

Protective painting

The methods already described to limit corrosion are normally supplemented by painting. The method of painting normally used is to choose a paint containing an inhibitive pigment for the lowest coat or primer. The primer is then covered with two or more outer coats to protect the primer from chemical change and mechanical damage.

In this section only the primer used in the painting of metals will be mentioned; a more general discussion of the nature of paints was dealt with in Chapter Eight.

The protection provided by the paint is due to:

(a) The presence of constituents called inhibitive pigments which suppress the anodic and cathodic reactions.

(b) The slowing down of the movement of ions through the coat.

Some cathodic protection is provided by metal pigmented paints.

Red lead in linseed oil

This is a well-known priming paint which is an effective corrosion inhibitor. The inhibitive action is complex but one part of the explanation is possibly the reaction between red lead and the extremely corrosive sulphur dioxide found in urban and industrial atmospheres to form lead sulphate. Non-oil-based red lead primers are available which dry quicker than the traditional primer and are reputed to give good protection uncoated for long periods.

Chromate metal primer

This is used for iron, steel, and aluminium. It is lighter coloured than the red lead primers and can be used under pale coloured finishes. It is suitable for composite articles since it can be used on wood and most metals. The inhibiting properties of zinc chromate are possible due to the formation of insoluble zinc hydroxide and the formation of a surface film of iron and chromium oxides which restricts the passage of ions from the metal to the solution.

Calcium plumbate metal primer

This is designed for use on new untreated galvanised iron. It can be used on wood as well as galvanised iron and is a suitable treatment for composite articles such as window frames. It has been suggested that the interaction of products formed by decomposition of the plumbate and dissolution of iron at the anodes, effectively stops the anodic process.

Etching or wash primers

These are usually supplied as two-solution paints. The one solution contains phosphoric acid, the other may consist of zinc chromate pigment in resin. The two are mixed just before application. When applied to the metal they rapidly set to a solid film. They are most satisfactory for use on aluminium and possibly steel. This is claimed to be a most effective method of pre-treating aluminium on building sites and it provides an excellent surface for the subsequent coat of metal primer, which should be applied within 24 hours because the etching primer is sensitive to water.

Tar and bitumen paints
These may be based on coal tar products, oil refinery products, or natural bitumen. They are used in industrial areas where oil based paints deteriorate rapidly. One problem is to produce a black composition which does not soften and run when it gets hot, or crack when it gets cold. They can be made water-repellent, and thick coatings can be applied. It appears that in some of these paints unidentified substances act as inhibitors and provide protection even when the steel is scratched bare.

Questions
1   When a metal corrodes, what is normally the end product?
2   Which metals corrode most readily and which corrode least readily?
3   (a) What is the 'electrochemical series'?
     (b) Arrange the following metals in their correct order in the series—Gold, Aluminium, Zinc, Lead and Iron.
4   State the meaning of the following terms:
(a)   anode
(b)   cathode
(c)   electrolyte.
5   Describe fully the corrosion of iron under:
(a)   acid conditions
(b)   neutral conditions.
6   How does differential oxygen concentration cause corrosion of:
(a)   mild steel
(b)   stainless steel?
7   What is anaerobic corrosion? How may it be limited?
8   Explain the difference between the techniques of (a) anodic and (b) cathodic protection. Compare the effectiveness of these methods.
9   The corrosion of iron and steel represents a tremendous waste of natural resources. Outline what steps could be taken to limit this corrosion in, for example, the motor car.
10 Most methods of protecting metals against corrosion are supplemented by painting.
(a)   Explain how the coat of primer gives added protection and name two primers in common use.
(b)   Why is the coat of primer normally painted over?

Ten   The Supply and Treatment of Water

The purity of water for domestic consumption is a vital factor in maintaining public health. Water which contains harmful germs or minerals can cause disease and illness for large numbers of people. The seas cover roughly two-thirds of the earth's surface and are a vast source of water. Sea-water is, however, unfit to drink because it contains a high proportion of minerals and salts which have been washed out of the soil.

The water cycle

Water evaporates from the surface of the sea and is carried over the land by winds. The clouds which form are then cooled as they rise over mountains, the water vapour condenses and returns to the ground as rain (Fig 55). Rainwater is, initially, free of foreign matter. However, as it falls through the air it dissolves substances such as oxygen and carbon dioxide. In industrial areas it will

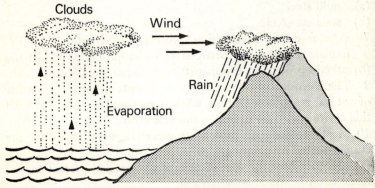

Fig 55

dissolve additional sulphur dioxide plus dust and germs. As it trickles over the surface of the earth or sinks down through the soil, it dissolves substances from the soil and the rocks. In addition it may dissolve animal or vegetable refuse.

Water supply

Water is obtained for domestic and industrial use either from reservoirs, which receive the water from upland catchment areas, or from rivers and wells. Whichever the source the water has to be purified before it is used.

Firstly the fairly coarse impurities in the water are removed by passing it through a primary filter. The primary filter may consist of a concrete basin with a layer of sand two feet thick supported on a bed of gravel. This initial filtration does little to remove bacteria or fine suspended matter. A second filtration is therefore necessary. Two processes are in use:

(a) *Slow sand filtration*
Water is allowed to flow slowly through beds of fine sharp sand. A biological layer forms on the sand which destroys the bacteria. It also filters out the very fine particles suspended in the water. In due course the organic layer grows to such an extent that filtration becomes too slow. The top layer of sand is then removed and cleaned. The disadvantage of this process is that it is too slow to cope with the demand for water from modern cities and industry.

(b) *The rapid sand filtration process*
Higher pressures are used to force the water through the sand filter bed which is sometimes contained in enclosed steel filter vessels. The sand is purified by forcing water upwards through the sand.

This is a less efficient process than the slow process and to assist with the removal of the fine particles and bacteria chemical coagulants are added. These are chemicals such as aluminium sulphate plus *modified* sodium silicate. When added to the water they form a curd-like precipitate. This settles to the bottom of the tank carrying the impurities with it. The water is first coagulated in this way and then filtered. Finally the water is sterilised by the addition of chlorine. A fairly recent sterilisation technique is to give the water a comparatively large dose of chlorine which disinfects it quickly.

The water is allowed to pass through a contact tank which removes the excess chlorine. The water is then sterile and tasteless.

Water treated in this way is safe and fit for domestic use but it will still contain dissolved salts and minerals which can cause the water to be chemically *hard*. This can be a serious disadvantage in certain cases.

Hardness of water

As it falls, rain dissolves carbon dioxide from the air to form carbonic acid H_2CO_3. The dissolving power or solvent action of the water is thereby increased. If the rainwater comes into contact with calcium carbonate $CaCO_3$ in the form of limestone and chalk, it dissolves it, forming calcium bicarbonate in solution. Similarly it dissolves magnesium carbonate which occurs naturally as magnesite and dolomite—

$$CaCO_3 + H_2CO_3 \longrightarrow Ca(HCO_3)_2$$
$$\text{Calcium bicarbonate}$$
$$MgCO_3 + H_2CO_3 \longrightarrow Mg(HCO_3)_2$$
$$\text{Magnesium bicarbonate}$$

Water also dissolves many other compounds from rocks with which it comes into contact. The most important are the sulphates, chlorides, and nitrates of magnesium and calcium. These, together with the bicarbonates, are the most important of the impurities in natural water.

Other impurities which are found are:

(a)   The salts of sodium; these are not objectionable unless they are present in large amounts.

(b)   Small amounts of silica SiO_2 may also be present; this can be a nuisance if the water is to be used as a boiler feed.

(c)   Iron may also be present as ferrous bicarbonate $Fe(HCO_3)_2$; for certain purposes this must also be removed.

(d)   In moorland districts the water is often acidic due to the presence of organic acids from decaying vegetable matter. The acids will need to be neutralised in order to prevent corrosion of lead and iron and to make it suitable for industrial use and domestic consumption.

A hard water is one which does not easily give a lather with soap; a soft water is one which lathers readily with soap. Hard water fails to give a lather with soap because the salts dissolved in the water react with the soap to give a useless curdy precipitate or scum. It is usual to classify the hardness as being of two types:

(a) *Alkaline or temporary hardness*

This is caused by the presence of calcium or magnesium bicarbonate and it can be removed by boiling the water. When the water is boiled the bicarbonates are converted to the corresponding carbonates—

$$Ca(HCO_3)_2 \longrightarrow CaCO_3 + H_2O + CO_2$$
$$\text{Calcium}$$
$$\text{carbonate}$$
$$Mg(HCO_3)_2 \longrightarrow MgCO_3 + H_2O + CO_2$$
$$\text{Magnesium}$$
$$\text{carbonate}$$

The carbonates are precipitated from the water and, in the case of the calcium salts, there is almost complete removal of the hardness. In the case of the magnesium salts the hardness is only partially removed.

(b) *Non-alkaline or permanent hardness*

This is not removed when the water is boiled. It is caused by the presence of salts such as chlorides, nitrates, and sulphates, of calcium and magnesium which are not affected by boiling. Both types of hardness can be removed by chemical treatment which results in the removal of calcium and magnesium salts as compounds of low solubility. Consideration of some of the problems outlined below will indicate why it is often necessary to remove the causes of hardness.

Hard-water problems

(a)   Waste of soap. Calcium and magnesium bicarbonates, chlorides, and sulphates, form an insoluble scum of calcium or magnesium soap. Until all the salts have reacted the soap is unable to do its job and is being wasted. The scum spoils the finish of fabrics in laundries and textile works.

(b)   It causes deposits of scale on the inside of boilers. The scale consists mainly of insoluble calcium bicarbonate originally present in the water. This scale reduces the conductivity through the boiler tubes thereby reducing the efficiency of the boiler.

(c)   Dissolved salts such as calcium sulphate are not decomposed by boiling. They will accumulate in the boiler water until they become so concentrated that they crystallise out on to the boiler metal.

(d)   Hard water can, for example, interfere in such processes as dyeing (by reaction with the dye) or in tanneries (by reaction with the tanning chemicals).

Water-softening processes

Precipitation methods

(a)   *The lime-soda process*
The water is treated with sodium carbonate and hydrated lime (calcium hydroxide). The calcium salts in the water are precipitated as calcium carbonate and the magnesium salts as magnesium hydroxide.

$$Ca(HCO_3)_2 + Ca(OH)_2 \longrightarrow 2\ CaCO_3 + 2\ H_2O$$

$$Mg(HCO_3)_2 + 2\ Ca(OH)_2 \longrightarrow 2\ CaCO_3 + Mg(OH)_2 + 2\ H_2O$$

$$CaSO_4 + Na_2CO_3 \longrightarrow CaCO_3 + Na_2SO_4$$
Calcium
sulphate

$$MgSO_4 + Ca(OH)_2 \longrightarrow Mg(OH)_2 + CaSO_4$$
Magnesium   Magnesium
sulphate   hydroxide

The magnesium salts are removed as magnesium hydroxide which is less soluble than magnesium carbonate. The calcium sulphate formed during the last of the above reactions will be precipitated by reaction with the sodium carbonate. The mixture of carbonates and hydroxides settles out in a tank. The softened water is finally filtered through a sand filter.

(b)   *Caustic soda and caustic soda—sodium carbonate process*
With some waters the precipitation of calcium carbonate and magnesium hydroxide can be effectively brought about by the use of caustic soda (sodium hydroxide) alone or in conjunction with sodium carbonate.

(c)   *Sodium aluminate*
This is often used in conjunction with lime and sodium carbonate or caustic soda and sodium carbonate in modern practice.

Exchange Methods

Base exchange or zeolite process

The materials used for this process were natural compounds consisting essentially of sodium aluminium silicate. Such naturally occurring minerals are called *zeolites*.

Synthetic organic materials called *ion exchange resins* have now been developed and are used in place of the zeolites. Artificial zeolites are, however, less durable and more easily affected by acids than natural zeolites.

Action : The calcium or magnesium compounds present in the water are replaced by the equivalent amount of the corresponding sodium compound. The sodium salts which are soluble pass out with the water leaving the exchange unit,

e.g. $CaSO_4 + Na_2X \longrightarrow CaX + Na_2SO_4$
        Zeolite
      $Mg(HCO_3)_2 + Na_2X \longrightarrow MgX + 2\,NaHCO_3$

Regeneration

When the exchange capacity is exhausted regeneration of the zeolite is carried out by passing a strong solution of brine (sodium chloride) through the zeolite bed. Sodium is replaced in the zeolite, the solution of calcium and magnesium chlorides passes out to drain, and the cycle can be restarted,

e.g. $CaX + 2\,NaCl \longrightarrow Na_2X + CaCl_2$
                    Regenerated zeolite

Advantages

The main advantages of the process are that:
(a) There is no sludge to be disposed of.
(b) It is simple to operate.
(c) Waters of low hardness can be softened more satisfactorily by this method than by the precipitation methods.

Disadvantages

(a) A base exchange softened water is generally more corrosive to iron and steel than the original unsoftened water.
(b) A considerable amount of water is wasted in regeneration.

(c)   The concentration of dissolved salts in the water is not reduced by this process.

(d)   Suspended impurities must be removed from the water beforehand to prevent clogging of the bed.

(e)   Certain dissolved substances, such as iron, must also be removed beforehand because they adversely affect the zeolite.

Demineralisation softening process

Synthetic materials have been produced which have the property of exchanging anions (negatively charged atoms or groups of atoms) such as chloride Cl^-, nitrate NO^-_3, and sulphate SO_4^{2-} ions, for hydroxyl ions (OH^-). They are called Anion exchange materials. They are normally used together with a material which exchanges hydrogen ions H^+ for the calcium Ca^{2+}, magnesium Mg^{2+}, and sodium Na^+ ions in solution. This is called a *cation exchange material* (a cation is a positively charged atom or group of atoms). The result is water, which is substantially free from dissolved salts, which compares with distilled water for purity.

The treatment of acid waters

Moorland waters are normally acid, due to the presence of uncombined carbon dioxide or organic acids formed as a result of vegetable decomposition. These waters are normally soft but they do have a corrosive action on iron and lead. To prevent this happening the water is treated with calcium hydroxide, sodium carbonate, or sodium hydroxide. If the correct quantities are added the waters may be made sufficiently alkaline to prevent the corrosion of iron and lead taking place.

Plumbo solvency

The solvent action of water on lead is very important in the supply of drinking water because serious lead poisoning may result. At any given time the water will only contain a small amount of dissolved lead; but this lead accumulates in the body and in time the level builds up to produce lead poisoning. As has already been mentioned, it can be prevented by treating the water with calcium hydroxide. A characteristic symptom of lead poisoning is a blue line on the edge of the gums.

The corrosion of iron
Iron corrodes most rapidly when in contact with acid water. It is essential that in all types of heating systems the water should be neutralised, and it is normally made alkaline. This is done by the addition of alkali or chromate, etc. The nature of the substance added is decided by the circumstances of each case.

EXPERIMENT 38
To illustrate the action of soft water on lead (plumbo solvency).

Apparatus
Two 250 ml beakers, two pieces of clean lead foil, soft water (distilled), hard water.

Procedure
Stand a piece of lead in each beaker, partially cover one piece with soft water and the other with hard water. (If necessary make hard water by bubbling carbon dioxide through lime water.) Leave the two beakers for several hours. The soft water becomes cloudy or turgid but the hard water remains clear. Now bubble hydrogen sulphide gas through each of the solutions. The distilled water turns chocolate-brown in colour, owing to the formation of lead sulphide, whilst the other remains clear.

Estimation of the hardness of water

The hardness of water is usually expressed in parts of calcium carbonate, equivalent to the calcium and magnesium salts per million parts of water.

Parts per million is abbreviated to p.p.m.

EXPERIMENT 39
Estimation of the hardness of water.

A   *Estimation of the temporary hardness*
Apparatus
Burette, 100 ml graduated flask, hard-water sample, M/10 sodium hydroxide, M/10 sodium hydroxide, M/10 potassium palmitate, methyl orange and phenolphthalein indicators.

Procedure

Measure out 100 ml of hard water in the graduated flask, and transfer it to a conical flask. Add one drop of methyl orange as indicator. Fill the burette with the M/10 nitric acid and run it into the flask with constant shaking, until the addition of one further drop of nitric acid causes a permanent change of colour in the flask (from colourless to red). Note the volume of acid added; let it be a ml.

Equation

$$Ca(HCO_3)_2 + 2 HNO_3 \longrightarrow Ca(NO_3)_2 + 2 H_2O + 2 CO_2$$

From this equation it can be calculated that temporary hardness of sample = 50 a p.p.m. of $CaCO_3$.

B   *Estimation of the total hardness*

The solution from Experiment 38 is not softened because calcium bicarbonate is replaced by calcium nitrate. Boil for five minutes to drive off the carbon dioxide, cool, and add one or two drops of M/10 sodium hydroxide until the methyl orange becomes colourless.

Now add a few drops of phenolphthalein indicator. Having rinsed the burette thoroughly fill it with potassium palmitate solution and run it into the flask until a further drop produces a permanent pink colour in the water.

$$CaSO_4 + 2 K.P \longrightarrow CaP_2 + K_2SO_4$$
   Potassium   Calcium
   palmitate   palmitate

where P represents the palmitate group.

Calcium palmitate is insoluble and is precipitated. When this precipitation is complete an excess of hydroxyl ions is produced in solution causing the phenolphthalein to turn pink.

Let amount of potassium palmitate = b ml

Total hardness = 50 b p.p.m. of $CaCO_3$.

C   *To estimate the permanent hardness*

This is the difference between the total and the temporary hardness.

Therefore permanent hardness = $50(b - a)$ p.p.m. of $CaCO_3$.

Questions

1 What is meant by the water cycle?

2 Name two gases which dissolve in rainwater as it falls through the air.

3 State one advantage and one disadvantage of the rapid sand filtration process.

4 What is meant by the terms 'hard water' and 'soft water'?

5 Explain the difference between permanent and temporary hardness.

6 What are the problems which arise from the use of hard water?

7 Compare the base exchange process for water softening with a precipitation method. State two advantages and two disadvantages of each process.

8 List the following in order of purity:

  rainwater
  tap water
  river water
  demineralised water.

9 Explain what is meant by plumbo solvency.

10 Why is sea-water not fit for human consumption? Suggest how it might be purified. What problems would be solved by large-scale purification of sea-water?

Appendix

Questions for Chapter one

1   Give an account of the manufacture, characteristics and uses of the following types of lime:
(a)   quicklime,
(b)   hydrated lime,
(c)   lime putty.

2   Differentiate between:
(a)   hydraulic lime,
(b)   non-hydraulic lime,
(c)   magnesian lime.

3   List the British Standard tests applicable to building limes.

4   Describe, with sketches, the test for the workability of a lime putty.

5   (a) (i) What is the cause of unsoundness in hydrated lime.
        (ii) What defects may arise due to unsoundness?
     (b) Describe, with sketches, the tests for soundness.

6   (a) What types of lime require to be tested for hydraulic strength?
     (b) Describe, with sketches, the hydraulic strength test and indicate how the result is calculated.

Questions for Chapter two

1   Write notes on *three* of the following types of gypsum plasters; plaster of Paris, retarded hemihydrate, anhydrous, Keene's, using the headings (i) manufacture, (ii) characteristics, (iii) uses.

2   (a) List the British Standard tests applicable to each class of gypsum plaster.
     (b) Describe the test for soundness.

3   (a) Describe, with sketches, the test for the transverse strength of a Class B gypsum plaster.

(b) To which other class of plaster does this test apply and how is it modified in this cases?

4 (a) Describe, with sketches, the test for the mechanical resistance of a Class B gypsum plaster.

(b) To which other classes of plaster does this test apply and how is it modified in these cases?

5 Compare the setting and hardening action of a gypsum plaster with that of a lime plaster giving the appropriate chemical equation(s).

6 What precautions should be taken when mixing gypsum plasters with other materials?

7 Write an account of *three* possible defects which may arise in gypsum plasterwork and suggest remedies.

Questions for Chapter six

1 Explain, briefly, how the rocks in each of the three main geological classes have been formed.

2 (a) Give *one* example of an igneous rock used in building.

(b) State (i) the group to which this rock belongs,
           (ii) where it is most commonly found,
           (iii) one application in building.

3 To which class do the following belong:
(i)   marble,
(ii) sandstone,
(iii) granite,
(iv) slate,
(v)   limestone?

4 (a) List the constructional uses of the materials given in Question 3.

(b) Justify the list by a consideration of the composition and physical properties of the materials.

5 Write notes on the factors which influence the rate of weathering of building stone.

**Books are to be returned on or before
the last date below**